Idaho Falls

IDAHO FALLS

The Untold Story of America's
First Nuclear Accident

Published by ECW Press
2120 Queen Street East, Suite 200,
Toronto, Ontario, Canada M4E 1E2
416.694.3348 / info@ecwpress.com

NATIONAL LIBRARY OF CANADA CATALOGUING IN PUBLICATION DATA

McKeown, William
Idaho Falls : the untold story of America's first nuclear accident /
William McKeown.

ISBN 978-1-55022-562-4
ALSO ISSUED AS:
978-1-55490-562-1 (PDF); 978-1-55490-543-0 (EPUB)

1. Nuclear accidents—Idaho—Idaho Falls—History.
2. Idaho Falls (Idaho)—History. I. Title.

TK1344.I2M32 2003   363.17'99'0979653   C2002-905423-0

Acquisition editor: Robert Lecker
Development editor: Jodi Lewchuk
Copy editor: Judy Phillips
Design and typesetting: Yolande Martel
Production: Emma McKay
Cover design: Rachel Ironstone
Front cover photo: John Rodriguez
All interior images appear courtesy of the U.S. Department of Energy

Printing: Marquis   10   9

This book is set in Dante and Blast

PRINTED AND BOUND IN CANADA

ECW PRESS
ecwpress.com

To my parents, James V. and Bonnie L. McKeown

Contents

Acknowledgments

I would like to thank the numerous people who shared their memories and their expertise about the SL-1 incident. Among those most generous with their time were John Byrnes, Robert and Bette Vallario, Stephen Hanauer, Dr. George Voelz, Don Petersen, Egon Lamprecht, Susan M. Stacy, Ed Fedol, and Diane Orr. Special thanks must also go to the graduates of the U.S. Army's Nuclear Power Program, who graciously cast their memories back more than forty years.

Nuclear theory, nuclear engineering and nuclear radiation are complex subjects, and I make no claim of expertise. In an effort to make this book readable—and to keep the focus on the human drama at the SL-1 reactor—I chose to boil these concepts down to the basics. Any errors in fact or context from doing so are entirely mine.

Finally, I would like to acknowledge my two wonderful daughters, Shannon and Caitlin, whose support never wavered during the long gestation of this book.

William McKeown
April 2003

Aerial view of the SL-1 testing station.

Prologue

It's hard to say when it all went to hell.

The mug shots of Jack Byrnes and Dick Legg taken at the beginning of their last military assignments betray no hint of the elemental forces that would soon engulf them. Missing from the muddy black-and-white photographs is evidence of the reckless passions and base instincts of two saboteurs—or, conversely, the bewildered innocence of a couple of patsies— that would make them the main players in one of the most mysterious human dramas in industrial history. Sure, if you search their faces for telltale signs of character, you might glimpse something intense and smoldering in the deep-set eyes of Jack Byrnes; you might detect a slight, smug smile on the squarish face of Dick Legg. But that's just mental rubbernecking. It's been more than four decades since the photographs were snapped, and they don't offer up much besides a musty smell and a record of bad haircuts. There's no indication the two weren't destined for long lives and ordinary deaths.

The passage of years and the death and silence of friends and family have left but the bleached bones of the two men's histories. They left no diaries, no record of notable achievements, and few anecdotes to hint at the mix of characteristics that made up their personalities. When they arrived in 1959 in Idaho's fertile Snake River Valley to take up what would be their final posts, Byrnes and Legg were still unformed, still works-in-progress, their individual potentials and futures as fuzzy as their service photographs. They were typical American boys on the cusp of manhood, at that age when character, talent, and limitations are just beginning to emerge. Born in the late 1930s in quiet American towns, the two played soldiers while real ones marched across Europe in World War II. As teens, they saw a great explosion in American power as those soldiers came home and rebuilt their country on the GI Bill, a package of government benefits that allowed soldiers to buy houses and attend college. Byrnes and Legg were young men of their time, schooled in the buoyant hopes of prosperity and order, confident that progress would give them better lives than their fathers, agitated by the sense of change and possibility that was sweeping America at the cusp of the 1960s.

The new god that emerged after World War II—the tripartite deity of industry, science, and technology—promised to make all these things possible, even for those of modest means and education. It was the atomic age, when Americans decided they had the knowledge, right, and wisdom to harness for other uses the terrible power released upon Hiroshima and Nagasaki. The atom had ended the war, smote the enemy. But surely its power could be tamed, used to power boats and planes, deliver unlimited electricity, revolutionize medicine. Atomic energy, America's leaders promised, would checkmate America's enemies while it bestowed on all others the gift of limitless energy. It was a seductive idea—using nondescript uranium ore to transform the world.

By the time Jack Byrnes and Dick Legg peered at the lens of a military camera, they had bought into this atomic-powered version of the American Dream. They were determined young men in a hurry, hungry for the good things in life, cocksure about their abilities and opportunities. For them and thousands of others, there was no better place to find the new America than at the National Reactor Testing Station, located on the vast expanse of sagebrush in Idaho's Lost River Desert. Established in 1949, the classified site lay just west of the Snake River Valley and was a monument to the golden age of nuclear science. It was a place where American men fearlessly played with the atom. When Byrnes and Legg arrived, almost two dozen nuclear reactors dotted the desert floor, prototypes of machines that would revolutionize propulsion and energy—and life. Government-issued films of the time celebrated the "new hope of the atomic era" and hinted at the blessings it would afford. Standing before models of prosperous and gleaming cities, film narrators—invariably white men in crew cuts and black suits—championed the glorious changes nuclear power would introduce to the arts, humanities, and sciences. An article in a 1958 issue of *National Geographic* concludes that "abundant energy released from the hearts of atoms promises a vastly different and better tomorrow for all mankind."

There were a few people who had doubts about the message. Ranchers and sheepherders in the American West were starting to voice concern about radioactive fallout from the nuclear weapons testing then being conducted in the Nevada desert, fallout that had made its way north to the Lost River Desert. Some experts and interest groups were raising questions about storing radioactive materials and locating nuclear plants near large cities. There were even some who scoffed at the rosy nuclear future portrayed in publications and documentaries at the time. But generally, the American people—and the folks in Idaho—believed the benefits of atomic energy outweighed the dangers it posed. For Byrnes and Legg, the lure of the atom had little to do with the promises and peril it posed to mankind. For them, it meant a paycheck and it meant a brighter personal future. They joined a select group of young engineers, construction workers, soldiers, and scientists flooding the wilds of Idaho to turn nuclear dreams into reality.

Then came January 3, 1961, a day that foreshadowed the dimming of the atomic dream, even if it remains a curiously obscure date to all but a few nuclear insiders. That afternoon, Byrnes arrived at a small experimental army nuclear reactor for his shift with Legg and a trainee, Richard McKinley. The three were scheduled to perform routine maintenance work. Nothing suggested that it would be anything other than an ordinary night.

But just a few short hours later, the ordinary became extraordinary. The events that unfolded in the crude silo of Stationary Low-Power Unit 1 on that January night would spawn more than four decades of scandalous rumors and speculation in the closed world of the nuclear industry. That one cataclysmic night would underscore the fragile line between the fallibility of man and the complexity of an intricate science. It would also reveal, but only much later, how a government shaped by a pervasive Cold War mentality would protect the then-fledging nuclear industry from public scrutiny.

While there were markers on the road leading to the chaos and calamity of that night—men with increasingly tumultuous personal affairs and a reactor with malfunctioning equipment—it was impossible to predict how these elements would collide in such a mysterious, unprecedented manner. Despite a nuclear testing facility that housed highly sensitive, top-secret equipment and some of the brightest minds of that generation, there was no way to measure, test, or imagine what would happen in the frigid southeastern Idaho desert on that January night. There was no way to predict the disaster.

1

Nuclear Apprenticeship

In late October 1959, United States Army soldier Jack Byrnes, twenty years old, set off from his hometown of Utica, New York. A trunk was tied to the roof of his black Oldsmobile; his wife, Arlene, was beside him in the front seat, and his son, Jackie, not yet two, was squeezed in the back among the couple's possessions. They were headed west to Idaho for a new adventure, a more promising future. A reel of eight-millimeter film the couple shot in Yellowstone National Park, not far from their new posting, snared them in celluloid. Jack is handsome and well built, his blond hair just starting to darken. Arlene, blinking at the camera, is thin, pretty, and vivacious. She dotes on young Jackie. Yellowstone is devoid of tourists. Old Faithful erupts on cue.

Born June 22, 1939, in Utica, John Byrnes III was the oldest of four children in a Catholic family. His father was a hard-working real estate salesman. By the early 1950s he was making a pretty good living—good enough to buy a cabin nestled amidst New York State's Finger Lakes. During summer and winter vacations, the elder Byrnes introduced his son—everyone called him Jack—to water sports and snow skiing, expensive pastimes even then. Soon Jack, a naturally athletic kid, was blasting watery arcs offshore from his dad's cabin and carving hairpin turns on the icy slopes of local ski hills. Bright and easily bored, the teenaged Byrnes didn't have a lot of interest in school. He liked girls, he liked driving fast, and he liked going out with his buddies to cruise Utica's hot spots and those in the nearby town of Rome.

"He was just a happy-go-lucky guy," recalls one acquaintance. "He was one of those daredevils. He'd try anything." Others, though, say the young Byrnes was more complicated than that, even in his adolescent years. Away from his friends, Jack was a serious, intense young man. He liked to do things his way, and when that didn't happen, his temper could flare.

During his high school years, he met Arlene Casier, who attended Rome Free Academy. Arlene, whose father had died, was living with her mother, a quiet, dignified woman. According to a good friend of theirs, both Arlene and Jack yearned for security and the good things in life, and both were in a hurry to get them. Byrnes, his father said years later, was a kid who wanted to grow up fast. At seventeen, he fudged his birth records and joined the United States Army. By the time he was nineteen, he had married his eighteen-year-old sweetheart and begotten a child.

After Jack's basic army training, Jack and Arlene plunged headfirst into the stressful world of military life, where the pay is low, moves are frequent, and extended family is always too far away. The couple quickly discovered just how little control they had over their new life. Their first posting in Newfoundland, a sea- and wind-battered province on Canada's east coast, could hardly have been more remote. During his stint in Canada, the Byrnes's first and only child, John—called Jackie by family members—was born. While adjusting to fatherhood, the young GI was assigned to mechanical training and spent his first year and a half in the military learning how machinery worked, how to maintain it, and how to fix it.

Sometime in 1958, Byrnes became aware of a new nuclear program run by the army at Fort Belvoir, Virginia, just south of Washington, DC. The army had built a small teaching reactor and was recruiting men from all three branches of the military service to learn the science of coaxing heat and electricity from atoms. The army planned to have the trainees operate a string of small, portable nuclear reactors that would be located in remote areas of the world. It seemed like an exciting prospect to Byrnes, certainly far better than fixing a tired generator at a bleak army base. He applied for the training program and was accepted. One of his classmates was Dick Legg, who had strolled into the nuclear world in much the same way.

*   *   *

Sailor Dick Legg, twenty-five years old, left for Idaho Falls, then a bucolic town with a population of about thirty thousand nestled on the banks of the Snake River, about the same time as Byrnes. The backseat of his car was filled with archery gear. No one sat in the passenger seat, but finding a girl would be a top priority after he settled in at his new posting. He, too, was excited about his move west. Although relocating to southeastern Idaho meant putting a lot of distance between him and his family in upper Michigan, Legg was an outdoorsman, and the landscape he saw flying past the car window as he drove confirmed that there would still be expanses of forests he could walk through with his bow at the ready.

Dick Legg was born in 1934, the youngest of Louis and Mary Legg's three sons. The family liked to say that the newest member of the family was the last of the three Ds: Don, Doug, and Dick. Louis Legg owned a small timber mill near the Huron National Forest. Legg and his brothers were bused, and later drove themselves, from their hometown of Roscommon, population five hundred, north to the larger town of Grayling to attend school. Legg, who showed no interest in sports, was a B and C student; smart enough, but distinguished mostly by his classroom antics, which earned him a reputation of being a class clown and a prankster. Legg's favorite way to spend his free time was to take off on his own into the wilds of Michigan. What he really liked— long before he had a driver's license—was careening in one of his brothers' cars down country back roads. Before he had reached his teens, Dick had taken up archery, and it became a passion. He often roamed the woods near his home with a bow in hand. Something about the solitude of the sport agreed with him.

As a teenager, Legg worked at his father's mill during summer vacations, helping run a massive blade through the stripped logs. The work put muscles on Dick, a source of pride for a guy who was touchy about his height of five foot six. But even with the muscles, Legg didn't easily attract women; he sported black-framed "geek" glasses, a pouty lower lip, and a large mole near his left nostril. But he did get attention, both wanted and unwanted, with his quick wit, smart-ass comments, and pranks. Legg's cousin remembers him as a jokester with a ready smile and, as the youngest in the family, a kid not weighed down by the expectations attached to his older brothers. The only major shadow over Dick Legg's young life was cast in 1949, when he was fifteen. His brother Doug was driving a new car down a back road at high speed when he lost control of the vehicle, crashed, and was killed.

After Dick graduated—with no great distinction—from high school in the mid-1950s, he drove to Grayling to enlist in the navy. Attending college apparently was never considered; there were no higher education degrees among the Legg men. But Dick's oldest brother, Don, had been in the service, and it seemed like a good place to learn a trade. Dick settled on the SeaBees, the navy's construction battalions, and was slotted to become a construction electrician. For the next two years, he shuttled around the eastern seaboard from one training program to another, learning the intricacies of electricity— how to harness it and how to fix the machinery when something went wrong. Then Legg learned that the army was looking for sailors and airmen to join its nuclear program; it sounded interesting. Like Byrnes, Legg was ambitious and confident—cocky even—in his abilities, and he didn't see a great future in being a run-of-the-mill electrician.

* * *

In early 1959, Legg and Byrnes arrived at Fort Belvoir to begin the training course in the nuclear program. Although they would later prove to be a fateful pairing, the two men didn't appear to have interacted much during their training in Virginia. One classmate recalls that few of the men socialized, as there just wasn't enough time. Byrnes and

Legg, along with fifty-eight other men, were thrown into an intensive training course, with only four months devoted to academic courses before undertaking another four months of hands-on training in specialty duties at the fort's small training reactor, dubbed SM-1.

Ed Fedol, a fellow trainee at Fort Belvoir, recalls the facility's crash course on nuclear power: "We only had four months of academics: nuclear theory, nuclear engineering, some mechanical engineering, electrical engineering, radiological engineering," he says. "It was rather heavy. It was a tough four-month grind. Then you went through four months of specialty training. You were either going to be a mechanic, an electrician, an electronics technician, or a health physics person. After that, you went through four months of reactor operator training."

Fedol, like Byrnes and Legg, learned about the program somewhat incidentally. Like all the young men who enrolled in the Fort Belvoir training school, he had developed his own version of an atomic future, one that was invariably rosy: "I thought, 'This is a brand-new field, this is a way to get promoted.' I was twenty-five. I had done four years in the navy, from the age of seventeen to twenty-one. I was out for two years, and then I was broke and hungry and went into the army, and I had been in for two years. And here I'm thinking, 'This is a way to get promoted pretty quick.'"

Martin Daly, also a graduate of the program, says the young recruits might have lacked college degrees but they were subjected to the most intense, real-life training their instructors could devise. "The last part of the course, the operations phase, started at a control-room simulator located at the school," Daly recalls. "Yes, we had a full-blown nuclear power plant control-room simulator back in the '50s. The instructor had the ability to introduce any number of problems at any time and observe the reactions of the operator. We had to spend many hours in the simulator before we ever were allowed to sit at the controls of a real nuclear reactor. We also had to memorize the schematic piping diagram and electrical diagram of the entire power plant. We had to know the location and function of every valve, every pipe, and every device in the entire plant. When we finally got to work in the plant, we were assigned to work with experienced people as equipment operators. The equipment operator works in conjunction with the control room to see that the plant always ran smoothly and to correct any malfunction immediately.

"Finally, as a control room operator, we were put under enormous pressure to perform," Daly says. "Working with a live nuclear reactor, we had to perform cold startup procedures, system shutdowns, and recover from SCRAMS [emergency shutdowns]. We also had to learn to recover from loss of commercial power, synchronize with commercial power, deal with runaway turbines, and supervise wastewater treatment. And all of this while keeping an accurate log of every move we made while on shift."

Both Byrnes and Legg did well in training, though no one remembers them as standing out from any of the other men in the class. But their instructors decided that they had potential. Having passed the psychological test required of all trainees, the two, their bosses reckoned, were ready to take the next step and become licensed reactor operators. For that they would need to go to Idaho's Lost River Desert, to the National Reactor Testing Station, where the army had built a small nuclear reactor called Stationary Low-Power Unit 1, commonly referred to as SL-1.

*   *   *

Forty-one miles west of Idaho Falls, the Lost River Desert has always been one of those just-passing-through places in the American West. At the peak of summer, the sun and heat are relentless, and shade is as scarce as water. When the sun drops at dusk over the dry mountains to the west, the desert radiates an unsettling power as it gives up the heat of the day. In the dead of winter, the desert, assaulted by wind-driven snow, seems to stretch out endlessly.

Even fourteen thousand years ago, when the climate was cooler and the landscape was dotted with shallow lakes and forests, prehistoric humans didn't linger in this northeast corner of the Snake River Plain, a band of flat land that curls across southern Idaho like a crooked grin. Molten rhyolite bubbled up from the planet's core, flowed out of fissures like black tongues, and then retreated, creating tubes, or caves, of fantastic forms. Under the extreme pressure, the desert's surface cracked wide open and volcanoes pushed their way upward. Three mark the desert floor; the largest, Big Southern Butte, rises up almost two thousand feet.

The desert, implacable in its harshness, withstood the first incursions of the white man. Fur trappers passed through in 1818 but quickly decided the barren landscape harbored more hazards than pelts. In the 1840s, westbound emigrants crossed the desert on a cutoff from the Oregon Trail; still-visible wagon tracks that head toward the sunset indicate they didn't stay long. Gold and silver strikes in 1860 attracted miners to the mountains north of the desert, but they saw no riches in the sagebrush they traversed carrying their supplies.

By the 1880s, industrious Mormon farmers were flooding into southern Idaho as a result of increasingly crowded living conditions in Utah. They settled along the eastern and southern edge of the Snake River Plain. Here they could use the river to irrigate the crops of potatoes, sugar beets, seed peas, and wheat they grew in the region's light soil, which was enriched with volcanic ash and trace minerals. Other settlers—ranchers and sheepherders—claimed land in the mountainous valleys to the north, where the water flowed freely. By the early twentieth century, much of the Snake River Plain, from eastern Oregon to Yellowstone country, had been transformed into a green patchwork of farms and quiet Mormon towns. Just one scrappy town—folks called it Arco—had carved out a tenuous hold on the western fringe of the Lost River Desert. The town owed its dusty existence to only one thing: its residents—never more than a few hundred—who were adaptable. They'd moved the town three times since its founding in 1882, chasing whatever kind of fortune-seeker happened to be passing through at the time. But the desert itself remained untouched, its thorniness no enticement to humans. Until the military decided it was the perfect place to install a nuclear testing ground and practice blowing things up.

*   *   *

When the easterners—Jack Byrnes with his family and Dick Legg on his own—finally arrived in Idaho's Snake River Valley, they were awed. On the eastern horizon, Wyoming's Teton Range—at that distance barely an inch high—thrust skyward, jagged and one-dimensional. Only a wisp of cloud at the summit hinted at the winds that raked the range's granite flanks and loaded snow into cornices as menacing as a cocked gun. To the west of the valley, past the interlocking blocks of cropland and the Snake River, lay the Lost River Desert, a vast sweep of sagebrush and black lava beds pocked by the occasional crater. And over their heads was a sky they'd never seen before. By day, it was an endless sweep of delicate blue; at night, the stars glinted like stilettos. Under that sky, during that fall of 1959, a promising future seemed waiting to be claimed.

Idaho Falls, the settlement straddling the turbulent Snake River, took the arrival of outsiders like Byrnes and Legg in stride; it was used to change. Picturesque but plucky, it ebbed and flowed with the fortunes of its residents. Its resilience had caught the eye of concrete, steel, and lumber manufacturers, all of whom brought their industries to the area. So when men from across the country answered the call of the atom and descended on the city, the reaction of the locals was for the most part no more than a collective shrug. The presence of military personnel and scientists signaled just another transformation of a city that had gotten used to reinventing itself over the course of almost a century.

In 1865, freighter J.M. Taylor hatched a plan for transporting goods across the fast-moving Snake River. He oversaw the construction of a log toll bridge, a successful project that resulted in the settlement naming itself Taylor's Crossing in his honor. But the town's name changed a few years later to Eagle Rock, a moniker coined by a group of travelers who spied an eagle perched on a juniper tree growing on a large rock in the middle of the Snake River. But that name didn't last long either. The city settled on its current name of Idaho Falls on August 26, 1891. The name recognized the water-falls, almost fifteen hundred feet wide, that were created when William Walker Keefer built a dam and retaining wall to harness the power of the rough rapids that cut through the heart of the city.

The municipality chose to commemorate its vibrant his-tory with an official city seal that featured the most prominent symbols of life in Idaho Falls: a sun rising over snow-capped mountains, a long swath of field, a stretch of choppy river, an eagle with wings unfurled. The seal eventually came to include the symbol for atomic energy, a nod to the thousands of workers who, like Jack Byrnes and Dick Legg, settled in the region and made the daily trek into the desert to wrestle the atom into submission.

As the 1950s drew to a close, Idaho Falls was the quint-essential all-American town, a perfect place for a couple of young men to get their careers off the ground and begin raising families. While the Mormon presence grounded the town with its wholesome family values, the locals never had difficulty finding fun. Teenagers cruised the downtown streets, their car radios tuned to the latest hits being broad-cast on KOMA out of Oklahoma City. After dark, the daring ones would sneak out to Lincoln Avenue, just off the north highway and a favorite spot for drag racing.

When they weren't working, residents lined the sidewalks in front of the town's movie theaters to catch whatever Hollywood happened to be pushing. When the weather was cooperating, they strolled the greenbelt, the developed land skirting the banks of the Snake River, as well as the grounds of Tautphaus Park. Others preferred to drive to Peterson

Hill and look out onto the lights that flickered in the city as evening came.

Kids of all ages flocked to diners like the Arctic Circle and Doug's Dairyland and soda fountains like the one at the Don Wilson Drug Store. Local merchants kept the town outfitted in everything from work dungarees, plaid shirts, and work boots to the more fashionable items of the day. And there was a neighborhood grocer on just about every street corner, where residents could stop in for their twenty-cent loaves of bread and one-dollar jugs of milk.

Given its homegrown American charm, Idaho Falls seemed an unlikely posting for a young soldier and a sailor. There was no major base close by. No facilities to hone the arts of war. The town boasted an enormous Latter-day Saint temple instead of tattoo parlors and seedy bars. But what it did have was the National Reactor Testing Station. And for Jack Byrnes and Dick Legg, that made it the perfect town. They wanted careers in the fledgling atomic industry, and there could be no better place to get a decent foothold on that dream than at the sprawling, classified research site situated forty-one miles west of their new hometown, where more than two dozen experimental nuclear reactors were scattered throughout the desert sagebrush.

At this point—the early 1960s—the nuclear industry was still in its infancy. It seemed to offer unlimited opportunities, even for a couple of average guys from average backgrounds with average education. Everything was new. Everything was possible. And it was all being dreamed up, and built, at the Testing Station. The experience of working at the site would kick open countless doors for ambitious men like Jack Byrnes and Dick Legg. It was fitting that the two arrived in Idaho toting just a few possessions and a lot of big dreams. They were ready to explore the nuclear frontier; they wanted to be nuclear pioneers.

Just fourteen short months after their arrival at the National Reactor Testing Station, the two young and eager enlisted men would make atomic history. The fulfillment of their strange destinies would make them legends in the tight-lipped, insular world of the nuclear industry. The bits and pieces of their personal and professional lives would make them either the main suspects or the unfortunate innocents in one of the most bizarre stories never told to the American public.

2

Atomic Energy Meets
the Cold War

Following Japan's surprise attack on Pearl Harbor in 1941, the United States Navy decided it needed a place to test-fire behemoth guns before shipping them into battle. Officials thought the vast stretch of sagebrush and broken lava fields in southeastern Idaho just the place. The Lost River Desert, finally, would yield to man.

Late in the year, the navy withdrew from the public domain a swath of desert about nine miles wide and thirty-six miles long, laid a railroad spur to the middle of nowhere, and named the area the Naval Proving Ground. Soon, the crack of sixteen-inch battleship mortars could be heard reverberating across the desert, followed by the thud of shells crashing into the fallow earth. When the war ended and many of the great battleships were decommissioned, it appeared the desert would return to being uninhabited. But one of the last experiments on the Proving Ground, the classified Project Elsie, assured a continued presence. The project involved a different kind of sixteen-inch shell, one crafted of depleted uranium, a heavy metal that helped the projectile penetrate thick armor. But the obscure, silvery-white metal that gave the shell its density had other special properties, both wonderful and deadly.

Just a few short years after unleashing the destructive power of uranium on Japan, atomic pioneers in America were toying with the idea of directing the force contained within the uranium atom toward something more than a destructive end. Could the atom be controlled, played with, made to mambo for months, years even? If so, the implications for providing energy in one form or another were staggering. Atomic visionaries knew that the fission process, despite all its seeming complexity, came down to one simple principle: An atom creates energy—and lots of it—when it splits. The fission of one uranium atom produces ten million times more energy than the combustion of one carbon atom found inside a chunk of coal. Machines that could harness that power and convert it to electricity, trumpeted the popular press, would change the world by lighting homes, powering transport vehicles, and ushering in an atomic age of cheap electricity and Buck Rogers–like gadgets. The directors of America's new military nuclear program, the civilian Atomic Energy Commission (AEC), would need to oversee a barrage of experiments to give them a deeper understanding of nuclear energy before atoms could be lassoed for a purpose other than one big destructive bang. That meant they'd have to find a space to play atomic cowboy. The AEC, like the navy, decided the inhospitable Lost River Desert was a perfect place to do dangerous things.

On February 18, 1949, the agency took over the navy's

Proving Ground, added another two hundred thousand acres to it, and called it the National Reactor Testing Station (NRTS). The classified site, the AEC told wide-eyed and slightly nervous residents of southern Idaho, would lead the world in the development and refinement of nuclear reactors and the materials needed to make them run. The folks in the small towns ringing the eastern edge of the desert likely didn't grasp the subtleties of the technology. But they suspected it meant a wave of Uncle Sam's cash was heading their way, as well as jobs for themselves and an opportunity to sell baseball mitts, tires, and homes to the newcomers.

On August 24, 1951, in the shadow of Big Southern Butte, the Testing Station's first nuclear reactor went critical, that elusive moment when enough fissionable material is arranged in just the right way to achieve a sustainable chain reaction. Designed and built by the government's Argonne National Laboratory, Experimental Breeder Reactor No. 1 (EBR-1) was housed in a nondescript, boxy building at the end of a dirt road. It was the first nuclear reactor in the world to use enriched uranium and the first to use a liquid metal coolant to carry away the tremendous heat generated by a nuclear reaction. Six months after the first chain reaction, the scientists at EBR-1 set another first—and pulled off a crude public relations coup. By hooking the reactor to a small generator and lighting four incandescent bulbs hung by a wire, theirs was the first reactor to create a usable amount of electricity; it was soon providing all the electricity needed for routine operations. The feats were heralded in 1951 as proof that nuclear fission worked and could eventually be harnessed to produce large amounts of electricity. The accomplishments were noted on a chalkboard hanging near the reactor and were accompanied by the signatures of the sixteen men present when they occurred.

The EBR-1 reactor managed to chalk up yet another achievement, one with implications that sent ripples of excitement through the industry. After the reactor had been chugging away for more than a year, samples of the nonfissionable uranium 238 that surrounded the reactor—the more mundane cousin of the scarce uranium 235 that created the atomic reaction—were shipped to an eastern facility for testing. Those tests revealed that EBR-1 had woven gold from dross: it had turned the relatively useless U-238 into fissionable plutonium. A wave of enthusiasm swept the nascent community at the Testing Station as scientists envisioned a world powered by an unlimited power source.

But EBR-1 would also remind scientists that they were dealing with complex, elemental forces that didn't brook human error. In 1955, during a test that pushed the uranium fuel to extreme temperatures, an assistant reactor operator made a mistake. When the plant's power level reached the desired level of fifteen hundred kilowatts, he pushed a button that sent a slow-moving control rod into the reactor core instead of the faster one that would have immediately quenched the nuclear reaction. It took only two seconds for his superior to notice the mistake and hit the button for the correct rod, but it was long enough for half of the football-sized radioactive core to melt. Fifteen minutes later, radiation alarms sounded and the reactor was evacuated. The first unintended meltdown in American nuclear history was reported to the AEC—but not to the public. It would be another year before the news leaked out to mainstream media.

Soon there were other audacious experiments afoot in the desert, many aimed at answering fundamental questions of how to build nuclear reactors that were efficient, cost-effective, and safe. The second reactor at the site had a hundred holes punched into its shell, allowing scientists and engineers to subject all kinds of materials to a nuclear inferno. Materials thought potentially suitable for constructing the body and guts of reactors were bombarded with high heat, agitated neutrons, and intense radiation. Soon a series of reactors was being subjected to the most dangerous abuses and extreme conditions the nuclear trailblazers could dream up—just to see what would happen.

In 1953, a small reactor called BORAX 1 was installed inside a simple water tank whose top had been left open to the desert air. During the next fourteen months, scientists sitting in a remote trailer flogged it in every conceivable way. Time and again they yanked out the control rod that regulated the fission process, provoking dozens of "excursions," or sudden, sharp increases in the reactor's power level. The instantaneous fissioning of atoms would boil the water in the crude reactor, causing it to shoot up one hundred fifty feet into the open air—eruptions noticed by motorists traveling along the nearby public highway. Each time it was provoked, however, the reactor would shut itself down rather than creating an uncontrollable chain reaction of atoms—the dreaded meltdown in which radioactive particles and gases are spewed into the air. Those results heartened the driving force behind the reactor, Argonne Labs scientist Samuel Untermyer, who didn't think human or mechanical intervention was needed to stop a "runaway" reactor. He had hypothesized that a nuclear chain reaction in an overheated reactor core would eventually grind to a halt simply as a result of the air voids created by boiling water and steam, and the experiments conducted on BORAX 1 seemed to prove him right. In more than two hundred experiments, Untermyer's supposition that boiling-water reactors were "inherently safe"—a designation that represented the Holy Grail for reactor designers—seemed to be substantiated.

Still, Untermyer wondered if it were possible for these reactors to be pushed too far—if a radioactive core could melt and destroy itself—and he thought it would be instructive to prove that possibility. He calculated how much radiation might be released if BORAX 1 suffered a meltdown. He then sought permission from the AEC to destroy the reactor. He got it, with the condition that a meltdown could be initiated only if the wind was on a course away from populated areas. On July 22, 1954, the wind was blowing in the right direction. Operators in the remote trailer ejected the control rod quicker than it had ever been ejected before. It sent neutrons slamming into one another at an unprecedented rate. The reactor blew up almost instantly, with a force later estimated as comparable to three or four sticks of dynamite. A black column of smoke and radiation rose one hundred feet into the air. Two square miles of the surrounding desert were reportedly contaminated.

Onlookers had just witnessed the first meltdown in nuclear history. But it wouldn't be the last time that atoms would run amok in the Lost River Desert. Soon after the July 22 BORAX 1 meltdown, four reactors, all dubbed SPERT (special power excursion reactors), sprang up. Each had a different design and each tested different materials, but all four had one thing in common: they were built for the purpose of teaching the researchers what could go wrong in a nuclear reactor, how to avoid such situations, and just how dire the consequences would be if they couldn't. Years later, after fifty-two reactors had been erected on the desert floor—the most in any one place on earth—there had been twenty-seven meltdowns. Nine were intentional; sixteen were from pushing beyond the limits of technology or human knowledge.

More than one of these meltdowns were caused by Clay Condit, a thirty-one-year-old physicist working for Westinghouse Electric Corp., one of the first big nuclear contractors. Looking back, he's quite pleased about having played such a groundbreaking role in nuclear history. Now retired, he still looks every inch the stereotypical 1950s hero-physicist: tall and handsome, with flowing white hair and a stentorian voice. Condit says it's hard for young people jaded by high technology and cynical about nuclear energy to understand the intellectual excitement afoot in the desert fifty years ago.

"At that time, the site was called the National Reactor Testing Station, and there was a reason for that," Condit explains. "It was for testing. It was a playground. There was a lot of interesting stuff. It was fun. You could do anything. Nobody had been there before, you know. After we did tests, it was all unclassified, which was nice. So we'd go out and do the circuit of all the technical meetings and make presentations. And people would clap and ask you questions and want your opinions."

Susan Stacy, a historian and author of a book on the history of the Testing Station, says the collapse of the nuclear industry makes it hard to imagine the enthusiasm and passion among those who flocked to the Lost River Desert more than a half a century ago: "The spirit of patriotism was absolutely palpable. That's what motivated people—their patriotic commitment to the United States of America. To compare the excitement of those days with what's going on today is just enough to—oh, I don't know—turn your stomach. Back then, there was a lot of life. Reactors were running and water vapor was exiting the cooling towers. There was a hustle and bustle. And there were things going on that were cutting edge. They were winning the Cold War for the United States, and they were learning things and doing things that had never been done before. And that was just fabulous.

"After the war, all the scientists had proven was that they could blow up a bomb," Stacy continues. "They had not proven that a nuclear rector could be controlled and managed for the constant, safe output of electricity. They were just so far from that. There were so many unknowns in everything from material science to physics—the whole spectrum of things that were not known was very wide. And I think the people who were working in the field recognized it was terra incognita. This was the first time the continent had been landed on. I talked to some of the wives of the men who were scientists out there, the ones who, if the findings of a particular test were going to be available at 2 A.M., wanted to be there. That meant you didn't go home for dinner and, in fact, it meant you didn't go home at all that night. Well, how did these women feel about that? I had the impression that they shared the excitement. Even though their husbands were not necessarily able—because of security—to discuss the details of what they were doing, these wives seemed aware that the family was in this period of exploration and discovery. Maybe they were carefully hiding any resentments they may have had in the past. But if they had resentments, I did not detect them.

"I think there was a great deal of idealism about the potential for nuclear energy, this fissioning atom, to solve many of the world's energy problems. The vast divide between the rich and the poor, and rich nations and poor nations, really often boils down to a matter of scarce energy. In many countries of the world, human labor is doing the work. Idealistic Americans in the 1950s had some hope that the tedium and human misery associated with that kind of labor could be ended by putting the power of this little atom to use for the world's good."

And so, buoyed in part by such utopian visions, construction workers, soldiers and sailors, physicists and engineers swarmed southeastern Idaho throughout that decade. AEC buses ferried them to the Lost River Desert in three shifts a day from the neat little tract homes that were springing up all over Idaho Falls. There was a palpable sense of pride among the crew-cut men who disappeared into the desert for ten hours at a time, carrying their black tin lunch pails and slide rules. Using a nondescript mineral from the earth and enriching its powers, site workers designed and built contraptions that would soon power ships and submarines. They constructed prototypes of commercial nuclear reactors that promised to make electricity "too cheap to meter," a phrase nuclear proponents loved to bandy about. The dreamier of those slide-rule jockeys even believed that they were changing the world, engineering salvation through the table of elements.

By the late 1950s, millions of dollars were flowing to the Testing Station for the development of military projects

General view of the SL-1 facility.

generated by Cold War fears. Big commercial utilities were lobbying Congress to use the reactor technology developed at the Lost River Desert at proposed commercial plants near major urban areas. The engineers were confident that they had the nuclear dragon by the tail. And they were regularly— and usually safely—twisting that tail in the course of their desert experiments.

* * *

When Jack Byrnes and Dick Legg arrived in the Lost River Desert in late 1959, they must have been disappointed when the SL-1 reactor on the Testing Station's grounds came into view. SL-1 didn't look cutting edge. It didn't look like the kind of place where you twisted the tail of anything. It looked, in fact, like a grain operation. As they stepped off the AEC bus in the gravel parking lot and approached the chain-link security fence, they saw a two-story administration building and two long, pre-engineered "Butler buildings"—essentially war-surplus metal huts. On the eastern end of one of the huts was a three-story, thirty-nine-foot-wide by forty-eight-foot-high metal silo with a covered stairway twisting up its side to the second floor.

Buried in sixteen feet of compacted gravel and rock on the first floor of the silo was the reactor itself. Called a pressure vessel, it was a svelte fourteen-and-a-half-foot-tall container constructed of carbon steel, clad in stainless steel, and thermally insulated. Nestled inside a stainless steel cylinder, the vessel held a few pounds of enriched uranium, in the form of fuel plates, in its core. Tack-welded to the side of each plate was a long, thin strip of boron, a "poison" that absorbs neutrons and helped keep the vessel's chain reactions in check. Originally, scientists had planned to mix the boron within the uranium fuel itself, but they ran into development problems. Tacking the strips of boron to the fuel plates was a rushed solution—one that scientists would come to rue. The reactor was controlled by just five aluminum alloyed cadmium rods that were lifted out of or dropped into the two-foot-wide by three-foot-high core to excite or dampen the movement of neutrons within the uranium.

The control rods and motors that regulated their movement were located on the second level of the silo, in an area known as the reactor room. Also on the second story were the reactor's top shields—removable steel and Masonite plates that covered the vessel's head—and concrete shield blocks that could be removed to allow crews access to the control rods. Above the reactor room, on the third level, was an air-cooled condenser and fan room. The steam generated in the pressure vessel passed through a turbine located on one side of the reactor room's operating floor and was then carried to the condenser. The condenser returned the steam to its liquid form so that a feed pump could send it back into the reactor. This natural circulation system was regulated from the main instrument panel in the small control room attached to the back end of the metal building that abutted the silo.

The reactor began producing power in 1958. It was designed to create only a small amount of electricity, about enough to "heat the general's bath water," as one wag put it. Actually, SL-1 was merely a prototype for a series of portable reactors the army wanted to build in order to power military radar stations in the Arctic Circle. The radar stations, making up the Distant Early Warning (DEW) line, were designed to be America's first line of defense in detecting and tracking intercontinental ballistic missiles and Russian bombers should they cross the ice cap from bases in Siberia. And the army saw the small reactors as a way of snatching a bit of glory and money away from the navy, which was in the process of impressing Capitol Hill with its development of nuclear-powered submarines. The army reactors were going to be simple constructions, light enough to be airlifted to their destinations and as easy to put together as Erector sets. They would run continuously for three years on a single load of fuel; their above-ground structure eliminated the problem of having to sink foundations into frozen tundra; and their minimal water requirements were an added bonus. The reac-

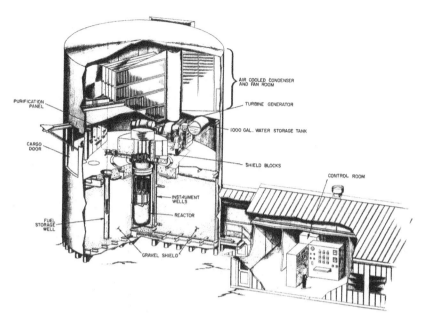

AIR COOLED CONDENSER AND FAN ROOM

PURIFICATION PANEL

TURBINE GENERATOR

1000 GAL. WATER STORAGE TANK

CARGO DOOR

SHIELD BLOCKS

CONTROL ROOM

INSTRUMENT WELLS

REACTOR

FUEL STORAGE WELL

GRAVEL SHIELD

SL-1 PLANT PERSPECTIVE

Interior of the SL-1 reactor.

tors' greatest advantage: they could be operated by just two or three men. Because the reactors were to be constructed in remote regions, no provision was made for a containment vessel, the thick concrete shell that surrounds most nuclear reactors and acts as a barrier against accidental releases of radiation. SL-1 was built to test the design of these portable reactors and to train the men needed to run them.

The SL-1 reactor was the smallest of the more than twenty in the Lost River Desert. It didn't break any new ground or explore any unusual technology. It was, perhaps, an interesting exercise in miniaturization and application, but that was about it. In fact, most of the four thousand people then working at the Testing Station were only dimly aware of the reactor, and that was only because it sat just three-quarters of a mile from the public highway they traveled to and from work.

Anyone assigned to the army's SL-1 quickly discovered that there was a definite hierarchy at the Testing Station, and that their little reactor was at the bottom of the prestige pole. Each branch of the US military had laid claim to a portion of the NRTS's desert site. In their designated areas, the army, navy, and air force were free to poke and prod the atom however they liked. As the 1960s approached, the navy's reactor program was premier. On March 30, 1953, a prototype reactor housed in a mock-up submarine that was sunk in a large basin of water and placed in the middle of the Idaho desert went critical. Less than two years later, the world's first nuclear-powered submarine, the USS *Nautilus,* cast off from a Connecticut boatyard, ushering in the nuclear navy and revolutionizing sea warfare. The navy program, with its dozens of topflight engineers and physicists and the cream of the naval officer crop, was the darling of the desert. The visionary behind the navy's top-notch operation was Rear Admiral Hyman Rickover, who was considered a demigod by many of the young engineers at the site because of his role in the development of nuclear energy.

When Rickover, a fastidious man and a consummate bureaucrat, visited the Testing Station, people snapped to attention. The son of a Polish-Jewish tailor from Chicago, Rickover quickly gained a fearsome reputation. He bullied and intimidated his Idaho naval staff and the young civilian engineers and physicists who worked on his projects; he insisted that every detail be disclosed to him. He demanded technical excellence and complete dedication to his vision. Years later, one engineer would recall how Rickover hated giving him Christmas Day off. He was universally regarded as an odd, driven man and—in the words of more than one nuclear veteran—a first-class son of a bitch. One wife of a civilian engineer later recalled a rumor about Rickover that had circulated during his reign in Idaho: it seems he had a penchant for stealing saltshakers from the homes where he was invited to dinner. Despite his eccentricities, Rickover got results. His reactor project, S1W, was a tight ship. "Rickover was a very strange man," recalls Condit, who worked for him. "I think he was incredible. He had a focus that very few people have."

If Rickover was incredible, Air Force Major General Donald Keim was apparently less so. He has the dubious distinction of being the earliest and most ardent supporter of perhaps the most ill-conceived idea in nuclear history: the nuclear-powered airplane. As early as 1947, the air force was predicting that it would take only five years to produce a bomber powered by a reactor. An elite circle that included the father of the atomic bomb, J. Robert Oppenheimer, thought the idea was sheer lunacy. Many others agreed. In 1948, a group of experts concluded it would cost at least a billion dollars and take fifteen years for scientists to solve the staggering theoretical and technical problems the project presented. And that wouldn't even address the obvious question: How much radiation would be released among civilians should a reactor-powered plane crash?

The air force and Keim were not dissuaded. The service was well aware of the abundant prestige and cash flow circulating in the nuclear business world, and it wasn't above playing on the country's Cold War paranoia in an attempt to claim its cut. Insiders began hinting that they suspected the Soviet Union was working on just such a plane. One skeptical congressman reportedly asked a high-ranking air force official if he wasn't worried about the prospect of radiation flying over the heads of Americans. The officer replied that yes, he was indeed worried. But, with a tortured logic that characterized much Cold War rationale, he added that he'd be even more concerned if the reactor-powered bomber circling above the heads of Americans wasn't American.

Eventually, the air force drummed up enough political support to overrule scientific misgivings about the project. In 1951, a civilian contractor was assigned the task of constructing the airplane reactor in the Lost River Desert. Four years later, the reactor went critical, proving atomic flight was at least a theoretical possibility. Although scientists had made no effort to shrink the reactor to a size and weight that could be lifted off the ground—it was huge—air force brass decided they needed a hangar for their hypothetical plane. In 1959, a massive hangar constructed amidst the sagebrush was unveiled. Measuring 320 feet by 234 feet and rising more than six stories, the hangar cost a cool eight million dollars. The air force planned to build an adjacent runway more than four miles long for its reactor plane, which designers thought would weigh 600,000 pounds and reach 205 feet in length. By comparison, the then state-of-the art B-52 bomber weighed 185,000 pounds, with a length of 159 feet.

But even after spending a billion dollars, the air force faced a problem: the reactor-powered bomber, if properly shielded to protect the crew from radiation, was simply going to be too heavy to lift off the ground. In what surely must have been an act of desperation, the air force briefly considered using less protective shielding and manning the cockpits with older men, ones who presumably would die a natural death before their insides were eaten away as a result of radiation poisoning. And there was another snag: the engines that were to be powered by the reactor had been spewing radiation into the sky during several years of test flights. Anti-nuclear activists later estimated that just one test of the nuclear aircraft engine had released 360,000 curies of radiation into the atmosphere. They pointed out that the Three Mile Island accident—which had scared the bejesus out of the American public—had released just 15 curies of radiation.

* * *

It was against this backdrop that SL-1 began its operational history. The reactor produced electricity in October 1958, two months after it first went critical, and in early 1959, it reached another milestone by generating power continuously for five hundred hours. But that didn't constitute much of a track record; it hadn't yet proved itself a triumph or a folly. Through the spring and summer of 1959, SL-1 ran fairly reliably, aided by the wits of the first generation of Army Nukes, as the graduates of Fort Belvoir had taken to calling themselves. The first operators at SL-1 were older men, well schooled in the workings of traditional power plants. Ed Fedol, who was a year behind Jack Byrnes and Dick Legg in the nuclear program, remembers the expertise they brought to SL-1: "I think the first class that went through [Fort Belvoir] in 1956 were all master sergeants, chiefs, e-7s. They were older guys, sixteen of them. What they [the selection committee] were looking for at the beginning were people with power plant experience. They could teach you the nuclear business, but they wanted people who were experienced in power plants.

If you had been an ex-navy machinist's mate working in the missile field or you did power plants aboard a ship, you were perfect—you were just what they wanted."

This first generation of reactor operators used their experience and ingenuity to handle the SL-1's initial teething pains, despite a lack of detailed policies and procedures. They also proved to be good instructors for the operator trainees who arrived at SL-1 in eight-month waves, each wave younger and more inexperienced than the previous. "In 1957, there was one class, and two classes each in 1958 and 1959. I was in the second class in 1959," Fedol recalls. "Each time they [the navy officers] started a new class, they were looking for younger people because they wanted someone who was going to be in the program for a long time. Those early ones weren't going to be around too long—most of them were going to be retiring."

Jack Byrnes and Dick Legg reported for duty at the Testing Station when the SL-1 reactor was a year and a half old. In the late fall of 1959, the two men were part of the fourth wave of trainees to descend on SL-1. By that time, conditions at the reactor had begun to deteriorate, but at a pace that made problems seem unrelated, and manageable. There were a few leaks of radioactive water. The odd seal was worn down. The control rods, crucial for mastery of the reactor, were starting to stick slightly as they were raised and lowered in the core. The boron metal placed in the pressure vessel to poison the reactivity of the uranium was starting to flake off and collect in a useless pile on the reactor vessel's floor. These last two problems were likely related to what the old hands considered a poorly designed reactor core. For a machine designed to be operated by only two men, there was an awful lot of personnel swarming around SL-1 just to keep it running properly.

People problems were also starting to crop up at the silo. A civilian contractor, Combustion Engineering Inc., had just taken over supervision of the day-to-day plant operations being carried out by the military and had yet to write and issue detailed policy and procedure manuals. Some of the veterans bemoaned the delay because they believed that the fourth wave of trainees, Byrnes and Legg's group, were undertrained and a little too unruly. They thought the new boys were in need of some direction and discipline. But Combustion Engineering supervisors were never asked to review the military's operation of the plant, and some said later they didn't suspect anything was askew.

Officially, the army was in charge of the soldiers, sailors, and airmen assigned to SL-1. But on the reactor floor, it was often difficult to tell who was ultimately responsible as inter-service rivalries developed among the men and their officers. And the local office of the AEC, which was overseeing the project, was taking a hands-off approach and remained distant. The cumulative effect of these management problems, it was determined later, led to lack of proper supervision and training for the new operator trainees.

"The army was just trying to build a reactor," recalls Condit, who was later asked to review the SL-1 project for the navy's Admiral Rickover. "Combustion Engineering was trying to patch it together to finish the project, and it was not a scientific effort. They were just running a reactor. The difference between the SL-1 and the navy projects was just unbelievable. There were less than a dozen army people assigned to that whole damn project, and they had a few

civilian guys from Combustion Engineering. And the AEC…
well, the AEC was not an influence out there."

It's unlikely that Byrnes and Legg were at all concerned
with what the more experienced Nukes thought was a pretty
low-dollar operation. From their perspective, they, along with
the ten other trainees who had come from Fort Belvoir, had
been thrown into a grueling routine. The men were assigned
to one of the three shifts that ran around the clock at SL-1,
and those shifts could change from week to week. The hours
were long; if the rides on AEC buses into and out of the
desert were included, the workday was at least ten hours.
The trainees were expected to hone their basic electrical or
mechanical skills; learn the complex procedures of operating
a reactor; and study for written and oral exams, administered
by a board of military and Combustion Engineering officials,
that would determine their promotions. Byrnes and Legg
found it challenging work, and they thought they were picking it up quickly.

*   *   *

If Jack Byrnes and Dick Legg were aware, during their first
months in Idaho, of technical complications and friction in
the ranks at their new assignment, they didn't say anything
about it on the record. Neither man seemed to be considering a long-term career in the military; both saw their stint in
Idaho as a means to a greater end: lucrative work at the commercial reactors they were hoping would pop up across the
country. If all went well in the Lost River Desert, they would
successfully complete their few months of training and pass
their operator exams. After six months of shift work as part

of a crew, they'd be eligible to undergo another set of assessments and climb the ladder to chief operator positions. With the experience they'd get supervising plant operations, they'd be shoo-ins for prime positions in the commercial world of nuclear power.

At the same time that the two men were finding their stride out in the desert, they were also starting to feel their way around their new hometown. Jack and Arlene had settled into a duplex on the east side of Idaho Falls. A number of their neighbors were also military folks, and Arlene found it comforting to be surrounded by people who too were newcomers to Idaho.

It didn't take long for the couple to establish a routine. Jack's long shifts at SL-1 meant many hours away from home, and Arlene spent her days running after the energetic young Jackie while trying to keep the house in order. It was nearly impossible to keep up with the toys that always ended up scattered over the carpet, the dishes that continually piled up in the sink, and the clothes that became dirty just as quickly as Arlene could clean them. She looked forward to the nights when Jack was at home and not too tired after she bathed Jackie and put him to bed; then they would have a chance to talk, a simple thing that seemed almost a luxury given Jack's shift hours.

When winter arrived, Jack's quiet time at home with his wife became even more rare. Since the day they had pulled into town, he had been itching to explore the nearby mountains. A dedicated skier, he found he'd landed halfway between Jackson Hole, Wyoming, and Sun Valley, Idaho—two of the best ski areas in North America. Even though his paltry army pay wasn't enough to fund trips to the hot spots with premier runs, he reveled in the powder and the steep slopes he found at the handful of small ski areas dotting the flanks of the lower Teton Range. The mom-and-pop establishments beat anything he'd skied in upper state New York, and they were only an hour's drive away from his home. That first winter in Idaho Falls, Jack's passion for the sport deepened, and he strapped his skis onto the roof of his Oldsmobile every free day he had. Something about the risk, the speed, and the solitary nature of skiing seemed to dovetail with his own character. He was happiest when his wooden skis were scribing long arcs into the side of a mountain.

Dick Legg, too, had discovered one of the charms of southern Idaho. Her name was Judith Cole, and she was a local girl who, just months earlier, had graduated from Idaho Falls High School. Like many young women of modest means and ambition in Idaho Falls, Judy had sought and landed a job with a contractor at the Testing Station. The jobs at the desert site, including Judy's position as a stenographer, paid better than most in Idaho Falls, but for many of the local girls, the money was just a bonus. The real prize was marriage to one of the young men who came to town to work at the NRTS. The girls saw them standing on the street corners in the morning—recent engineering graduates, soldiers, and sailors waiting for the blue government buses to whisk them westward into the desert. The men themselves may have come from towns just as small and slow-paced, but in Idaho Falls, the nuclear workers seemed exotic in the eyes of the young women who had spent their lives hemmed in by potato fields. "I know the girls around town quite liked all those service guys," said one resident years later. "They were somebody new, from someplace else. They were quite popular with the young girls." It was a phenomenon that irked the local boys.

The co-mingling of local girls and site workers had become so commonplace that when the funny and friendly eighteen-year-old Judy Cole began dating Dick Legg, by then twenty-six, no one raised an eyebrow. Except for her folks. Judy was still living in her parents' home on Capital Avenue, surrounded by the mementos of a safe, if a bit staid, childhood. Judy's parents liked Legg well enough, but they thought he was too old for their daughter, and she too young for a serious relationship.

Judy's family was Mormon—no surprise in Idaho Falls— and her parents may also have been disappointed that Legg wasn't. Nevertheless, it was a quiet parental reservation, and it didn't deter Judy from seeing more and more of Dick. With his bachelor digs close to the center of town, Legg occasionally stopped by the homes of his coworkers for a drink, but he spent most of his time courting Judy, taking her to movies or to dinner. If Judy saw in Dick security and the promise of a more exciting life, he saw someone who would defer to him. And Dick needed that. Those who got to know him found that the short, stocky Legg came with a swagger and a dash of arrogance. His personality quirks weren't glaring but were enough to occasionally draw attention. There's a story—third-hand at that—about how mismatched the two might have been. A graduate of the Army Nuke program, now a retiree living in Arizona, went through the Fort Belvoir training after Legg and Brynes and had met them only briefly. The nuclear veteran says a buddy recounted a conversation he had had with Legg:

"At Fort Belvoir, if you went up Highway 1 toward Alexandria, it's a typical army town. You've got bars, you've got honky-tonks, you've got hookers. It's the kind of life you lead sometimes if you're in the military. We called it The Strip; it's where all the action was. Legg told my friend, 'I've got it all figured out with these gals down on The Strip. I've got this first little bar down here. I sit at the bar, buy a drink. If I sit for a little while, a woman is going to come in by herself and sit at a table. I go over there and ask to buy her a drink. She either says yes or she says no. If she says no, I wait for the next one. If she says yes, I sit down and have a drink with her. Then I tell her, 'You know, this place has no class. I know a place down the road that has a nice band.' She either says yes or she says no. If she says no, I go back to the bar and wait for the next one. If she says yes, we get in the car and head down Highway 1. I pull in front of this drugstore a little ways down there. I open the door and say, 'Wait here, I have to go in and buy some condoms.' When I come back to the car and she's there, I know I have it made.' This was what Legg projected, that he was sort of a ladies' man—if there's any truth in it."

As the first six months in Idaho passed, Jack Byrnes's personality also came into sharper focus for his neighbors and colleagues. He was smart, no doubt about it, they said later. But he was young and a bit immature. He seemed to chafe at the responsibility of a family, remembers Stella Davis, the wife of a sergeant at SL-1 who lived across a courtyard from the Byrnes family.

Stella also noticed that Arlene seemed overwhelmed at times. She had married young. She had a toddler. She was living far from her childhood home. But Arlene's wasn't an uncommon story among the wives of the army's enlisted ranks. Stella and Arlene had known each other casually at Fort Belvoir, but the isolation of the Idaho posting soon drew them closer. Stella was a few years older than Arlene, and the younger woman often turned to her for advice. Stella had two children, one the same age as Arlene's Jackie. Arlene always had a multitude of questions about raising her son, keeping a home, trying to rein in her husband. Once a month, when their husbands got paid, Arlene and Stella would hire a babysitter for their kids and then drive east, halfway to the state capital of Boise, to shop for groceries at the PX, the military's own department store, at the Mountain Home Air Force Base.

"We were away from home," Stella recalls. "But when you're in the military, you're in a family. We helped each other. There were times when I needed a babysitter and she was there for me. And when she needed a babysitter, I was there for her. We hung out together because we had the two year olds and they played together. Arlene smiled a lot. She talked a lot. She and I always had good visits." Stella, though, couldn't help but notice that Jack wasn't around as much as he might have been. "He liked to go out and have a few beers with his friends."

Despite the tension brewing in Arlene and Jack's relationship, the spring of 1960 seemed a glorious time for all the young people who had come to work at SL-1. In March, Judy Cole, against the advice of her family, married Dick Legg, in a ceremony near Yellowstone Park. The newlyweds settled into their own place in Idaho Falls. That spring, Jack and Arlene Byrnes would get together with other "atomic" couples, easterners mostly. They'd buy beer and hold impromptu parties on the weekends, or they'd barbecue burgers at family cookouts. Occasionally the women would dress up and their husbands would take them downtown to The Rio movie theater for a Bob Hope comedy. Sometimes the group would pile the kids into the cars and drive out of town to fish along the Snake River or hike in the Tetons. No one had a lot of money, but it didn't seem to matter. They were all having fun on their days off, and the men were learning a trade at work that would eventually give them the lifestyles they dreamed of.

Byrnes and Legg didn't live and breathe the nuclear world as did their counterparts in Admiral Rickover's submarine program. They didn't toil under Rickover's strict rules and regulations. They weren't subjected to his tirades about dedication. Byrnes and Legg weren't scientists or engineers and likely only dimly grasped the cautionary lessons of the EBR-1 meltdown and the folly that was the airplane reactor program. The two newcomers took an enlisted man's view of the nuclear world: One foot in front of another. Just over the next hill. Pass the bottle. Maybe the atom was indeed a tyrant mistress with immutable qualities unforgiving of hubris, as their more experienced shift mates kept telling them. But to Byrnes and Legg, running their small SL-1 reactor seemed like no greater a task than turning up the stove gas under a kettle. And since they weren't Rickover's boys, they weren't subjected to his aphorisms—the pearls of wisdom he'd snarl out and expect his men to memorize, internalize, and heed—such as, "The whole reactor game hangs on a much more slender thread than most people are aware. There are a lot of things that can go wrong and it requires eternal vigilance."

It is more likely that Byrnes and Legg had in mind their

training at Fort Belvoir, where they were taught a much different lesson. Martin Daly, a graduate of the reactor program, remembers their gospel, perhaps based on Untermyer's experiments with the BORAX reactor in Idaho: "As anxious new students, we all asked our instructors about the safety of different types of power plants. The doctrine they preached was always, 'A boiling-water plant [like SL-1] can never blow up!' As the theory went, a boiling-water reactor was the safest of all reactors. We believed it. The designers, the officers, the instructors, and the students all believed it. Why not? After all, we were the pick of the litter. We were Nukes. We were career military people who believed in the United States of America. Our government would never lie to us."

3

"There Must Be Something Wrong at SL-1"

Despite the human tendency to analyze, to probe, to dissect, life often defies logic, resists being boxed up into neat packages. Talk of a "defining moment" is usually, in retrospect, just so much speculation. But for Jack Byrnes and Dick Legg, there was an event that did seem to mark some turning point in their lives. Though likely regarded as little more than a fairly typical incident at the time, people important in the nuclear world would later look back on the night of May 27, 1960, seven months after Byrnes and Legg arrived in the Lost River Desert for their great atomic adventure, and see the genesis of tragedy.

The evening began innocently enough, with Byrnes and Legg joining twenty-five other Nukes at the White Elephant Supper Club for a bachelor party for a civilian who worked at SL-1. A mix of men turned out that night: young servicemen, a sprinkling of officers, some civilian engineers, a few health physicists. Tables were pushed together for dinner. Afterward, the drinks flowed freely. Cigarettes started to pile high in the ashtrays. And, of course, there were the usual toasts, from the heartfelt to the crude, wishing the engaged man happiness in his upcoming marriage.

Later in the evening, as the party started to break up, a smaller group of men jumped into their cars and headed to the Boiler Room. The nightclub, in the basement of a building on Yellowstone Highway, featured strippers and as sleazy an atmosphere as you could find in a Mormon town. It was a favorite haunt of some of the young soldiers and sailors who worked at the Testing Station, and old-timers say it was tolerated by the conservative civic leaders. On that May night, the men pounded down drinks—tequila and whiskey mostly—as they watched the mostly imported talent bare their breasts. Sometime around 11 P.M., Jack Byrnes brought a young woman named Mitzi to the men's table, introduced her around, and encouraged her to join the group.

A half-hour before the bar's closing, Byrnes, Legg, and two SL-1 sergeants decided to move the party to the apartment of Byrnes's friend, a civilian employee at the reactor and one of the men gathered around Mitzi. Byrnes asked Mitzi if she'd like to come along; she agreed to. The six piled into the cars and headed to the apartment. The men were drunk and boisterous, excited by the woman in their midst. Sexual tension was in the air. But the party didn't last long. Close to 2 A.M., the building manager pounded on the door and told the group that residents were complaining about the noise. The men, intrigued by the flirtatious Mitzi and emboldened by the booze, decided they didn't want the night to end. They got into their cars again—even though by that time they were thoroughly drunk—and drove to the home of one of the sergeants.

Sitting in the living room of the small house near down-town Idaho Falls, the men gulped down more tequila and whiskey, a combination that surely lowered whatever inhibitions may have remained. Inevitably, their drunken banter turned to the subject of sex. Byrnes knew who Mitzi was, and the other men knew that proper young women didn't hang around strip clubs unaccompanied—at least not in Idaho Falls, and certainly not in 1960. An investigator who later had occasion to look into the evening of debauchery used a quaint phrase to describe Mitzi: "She proved to be a woman of easy virtue." Much later, stories would circulate that she was a hooker from Las Vegas who, each year, worked the small Mormon towns in Utah and Idaho. Details about Mitzi, her life, and her presence in Idaho Falls on that May night have never been explained in official documents. What *is* known? For one, she was a poor negotiator. At some point, with the men crowded around her in the living room, reeking of booze and seething with base passions, Mitzi offered to have sex with them for twenty dollars each. That was a lot of money for the military guys, who were lucky if they made four hundred dollars a month. Haggling ensued, and the price dropped, and kept dropping. In the end, Mitzi sold herself for two dollars a head. Some of the men accepted her offer, while others abstained. Jack Byrnes visited Mitzi in the back bedroom; Dick Legg didn't.

Sometime around 3 A.M., after Byrnes's quick assignation with Mitzi, he and Legg were standing in the living room behind a sofa, talking to one another. At that hour, and in the wake of the evening's events, the men were bleary-eyed and spent. Suddenly, the verbal exchange took a turn, and Byrnes and Legg started swinging at each other. Before either man could land a serious punch, one of their sergeants jumped up off the couch and pushed the drunken servicemen through the screen door and out onto the porch. The sergeant put himself between Byrnes and Legg before the fight could escalate and then pushed Legg toward his car. Byrnes walked back past the now broken screen door to a dying party. Afterward, neither Byrnes or Legg mentioned to coworkers what had provoked the scuffle. Later, when it became important, the sergeant who intervened between the two speculated that Legg made a snide comment to Byrnes about his liaison with Mitzi—either the morality or the brevity of it. Or maybe, he said, Byrnes had simply told Legg he was drinking more than his share of the communal whiskey.

It was a tawdry night, one of those embarrassing evenings best forgotten. But it wasn't. Soon, the drunken carousing would loom large in the personal histories and fates of Byrnes and Legg. The altercation may well have been simply drunken rivalry between two young men. It may have been a clash between two visions of morality. Or it may have held greater significance. In subsequent months, a rumor would surface about a tryst between one of the men and the other's wife. Had the mention of a liaison sparked a fistfight between the SL-1 coworkers that night? Could a fistfight over a woman—or women—have set up a deadly rivalry between the men, one that would color the history of nuclear energy? Only Byrnes and Legg could answer that question, and that wasn't going to happen. When the issue became important, a few would try to discover the motivation for the row. Others who were further removed would speculate and gossip. Even now, the reason remains elusive.

* * *

It's unclear what Byrnes thought of his actions that night. Maybe he regretted the evening—the hooker, the fight— and maybe he didn't. At the time of the bachelor party, Jack Byrnes was working part time at Kelley's Texaco Station on the Yellowstone Highway. He pumped gas, filled tires with air, wiped windshields clean of splattered spring bugs. He was the kind of guy who needed to keep busy; the job was easy, and he could pick shifts that worked with his schedule at SL-1.

According to one night operator at the Texaco station, Byrnes was well liked by the other gas jockeys; he fit in, he was just one of the boys. Except when his wife called the station. When that happened, it was as though a switch had been thrown in Byrnes. He invariably got angry and started to curse while talking with Arlene. His coworkers got the impression that Byrnes took the part-time job as much to get out of the house as to bring a few bucks home; they suspect- ed Byrnes was having marital problems. But even a second job didn't seem to put enough distance between Jack and his home life. One gas attendant recalled a night in the fall of 1960. Byrnes came by with a buddy and asked the attendant to cover for him by telling Arlene if she called that he was working that night but out on a call. She did phone—several times. Finally, at 2 A.M., the gas jockey cracked. He had to tell Arlene that her husband was out catting around.

Despite being stuck in the Mormon heartland, Byrnes had discovered a seamier side to life in the Snake River Valley. It wasn't in plain view—not like the strip of topless bars and tattoo parlors outside Fort Belvoir. But if you knew where to look, it was there. "There was a buttoned-up Mormon culture in Idaho Falls," says one longtime Idaho resident. "Yet Idaho Falls was the most notorious city around for bars and dives and prostitution and all the other comforts. When you got away from Sunday, there were six other days of the week."

One of Byrnes's friends says that by the fall of 1960, Jack was hitting local nightclubs once or twice a week—and Arlene didn't like it. Stuck at home during the day with little Jackie, Arlene thought the least Jack could do was take her out occasionally to dance, see a movie, eat a nice meal. When she complained loudly enough, Jack would grudgingly take her out for an evening. But he much preferred going out alone or with his friends. And when he was out on the town without his wife, one friend later told officials, Byrnes would often dance with unattached women he'd meet at the clubs.

A coworker at SL-1 remembered one particularly wild night. They were partying at a nightclub in Blackfoot, a small town thirty miles south of Idaho Falls. Byrnes decided sometime after 1 A.M. that he wanted to make last call for drinks at one of his favorite clubs in Idaho Falls, the Bon Villa. His friend couldn't talk Byrnes out of the notion; all he could do was hold on tight as Byrnes flung his Oldsmobile up the dark, two-lane Highway 91 at over one hundred miles an hour.

As snow blanketed the local ski hills in December 1960, Byrnes resumed his weekly treks to the slopes, sometimes alone, sometimes with buddies. He flitted from the mountains to the reactor to the Texaco Station to the nightclubs—anywhere it seemed but home. When he *was* at home, the fights between him and Arlene were increasingly frequent, and louder. The wife of one supervising military officer later said the rows had a certain script: Jack would criticize Arlene for not cleaning the house, and Arlene would retaliate by venting her frustration at Jack for rarely taking her out at night. The couple also often fought about money; Jack didn't think he saw enough of his paycheck. But it was Arlene who worried about the rent, the utilities, the food bills. Indeed, she had been forced to take a job at Newberry's, the five-and-dime store, to bring in some extra money. But even with that job and Jack's shifts at the Texaco Station, money always seemed tight.

A Testing Station worker who lived next door to the couple later told officials it seemed as though the two had a screaming match at least once a week; the noise carried through the duplex walls. Once the fights hit a certain frenzied state, Arlene had a habit of gathering up Jack's clothes and throwing them onto the lawn for all the neighbors to see. The wife of one of the SL-1 sergeants said the arguments became so routine that she felt compelled to talk to Arlene about it. The screaming matches, the older woman told her, were becoming the subject of gossip in the neighborhood, threatening to discredit the military folks who by then were being transferred to Idaho Falls in increasing numbers. Her words, she said, seemed to fall on deaf ears. The Byrnes's marriage, to outsiders at least, seemed to be crumbling. It wasn't hard to believe. Jack and Arlene had married young, having Jackie soon after. They had endured three moves in as many years. Perhaps they'd simply entered that stage where each was beginning to take the measure of the other, where romanticized dreams ran up against electric bills, dust on the furniture, and the same face every morning.

Arlene's friend Stella Davis lays the blame for the tension

in the marriage squarely on Jack's shoulders. He was handsome, bright, and personable when he wanted to be, she says. But he just wasn't able to own up to the responsibilities that came with having a demanding job, a wife, and a two-year-old son. He wanted to go out at night with his buddies, drink beer, and party. On his days off, all he wanted was to strap his skis onto his Oldsmobile and head into the mountains, where concentrating on the mechanics of skiing well would obliterate his worries and tensions. On the slopes, he could run as fast and wild as he wanted. If he did it right, he danced down the hill. If he didn't, he'd take a tumble, get up, brush himself off, and ride the lift back to the top to try it all over again. On the mountain, he was in control; he could do things his way. It wasn't like that down in the valley or out in the desert. "He was so young when he got married, and he was so young when he was a father," says Stella Davis. "It was too much. Take a nineteen- or twenty-year-old boy and he's not ready for marriage, or especially for being a father."

Things were also starting to unravel at the reactor. By December 1960, Byrnes's supervisors at SL-1 had judged him "not ready" for a promotion. They weren't thinking about his personal life, though many of the close-knit crew knew that Jack and Arlene weren't getting along. They'd heard snippets of angry bickering and the phone receiver being slammed down when Arlene called SL-1 while Jack was on shift. A few of the managers knew about the bachelor party, the prostitute, and the fight with Dick Legg. Coworkers had heard stories about Byrnes's carousing. Usually, none of that would have mattered to them; life in the young, enlisted ranks can be a rough-and-tumble world, after all. But Byrnes was proving as immature inside the reactor silo as he was in the outside world, and his supervisors were disinclined to trust him with controlling nuclear fission.

And they thought it a shame. Byrnes's supervisors recognized that the army specialist had above-average intelligence, that he was a cut above most of his colleagues. He was capable of doing good work. He was intuitive and curious, almost voraciously so. He wanted to know how the reactor worked; he was fascinated by the myriad systems needed to spark the atom, boil the water, spin the generator. In many ways, he was exactly the kind of young man the army wanted. But Byrnes had a few quirks when it came to his career. He disliked military authority—a real problem if you're a soldier. He was bullheaded and wanted to do things his way, an attitude that seemed presumptuous for a rookie. And he had an explosive temper, a potentially catastrophic characteristic for anyone routinely entrusted with radioactive uranium.

The supervisor of the reactor's maintenance crew, army sergeant Robert Bishop, later said that the hold on Byrnes's advancement was entirely justified: Byrnes was immature and wanted to move forward faster than his capabilities warranted. Furthermore, the cocky kid wasn't above playing office politics. Bishop recalled with some bitterness that Byrnes purposefully stirred up trouble between himself and the military superintendent of the plant, Richard Lewis. According to Bishop, Byrnes fed stories about his bosses—often with the details distorted—from one to the other. Byrnes's rumor campaign soured his own relations with Lewis and fueled the inter-service rivalry that was already plaguing the plant's operations. It wasn't until much later that Bishop and Lewis met for a talk and discovered that Byrnes was playing them against each other in the hopes of personal gain.

Byrnes was, in fact, establishing a reputation with all his superiors. Army sergeant Paul Conlon, the reactor's training officer, later told authorities that Byrnes "was a problem on the job...and was difficult to handle." He said Byrnes liked to work on his own and demanded full credit for what he accomplished, a corrosive attitude on a job that required teamwork. Most distressing of all, Conlon said, was that when Byrnes was displeased, everyone in the reactor building knew it. The young man regularly raised his voice and thought nothing of throwing tools around to demonstrate his displeasure.

That last shenanigan made an impression on army sergeant Gordon Stolla, who attended reactor school in Virginia with Byrnes and Legg and had been quickly promoted to chief operator at SL-1. He later recalled that Byrnes had moody spells a couple of times a week, during which he was morose, on edge, and angry. Everyone knew when Jack was having a bad day: in addition to chucking tools around the reactor building—a building filled with important pipes and gauges—his face would turn bright red. Byrnes was proving to be such a troublemaker that his immediate supervisors decided on a management technique likely unacceptable today. When Byrnes one day refused a direct order to make a scheduled check of the SL-1 plant, his immediate boss, chief reactor operator Sergeant Herb Kappel, got pugnacious. He offered to take the young soldier outside and "give him a lesson." Byrnes declined the offer and grudgingly performed the check. Kappel later told investigators that the plant's military supervisor had approved a bare-knuckled approach with the belligerent Byrnes.

Byrnes knew what his bosses thought, and it didn't please

him. He had moved west expecting to quickly earn his reactor operator's license and then a promotion to chief operator and shift supervisor. Four months after arriving in Idaho, he had passed his operator's test—almost everyone did. But the promotion to chief operator didn't seem to be in the cards. Byrnes had done good work on the mechanical crew, keeping the reactor up and running. He was smart and inquisitive, and so he had a hard time accepting that he wasn't moving up the ladder. He hated that he was going nowhere fast, while most of his Fort Belvoir classmates were already in supervisory positions. Being passed over for promotion was more than just embarrassing; It was a blow to his pride, especially since he knew he was as intelligent and capable as any of his cadre members. It also meant less pay. And a lack of supervisory experience could hamper his prospects of landing a job at one of the big commercial nuclear plants that everyone knew were soon coming to America. Byrnes complained to anyone who would listen that it just wasn't fair. As December 1960 wore on, Byrnes's resentment was palpable.

* * *

On the surface, Dick Legg's nuclear career was progressing somewhat better than Jack Byrnes's, which must have been galling to Byrnes. In September 1960, eleven months after Legg's arrival at SL-1, he passed examinations administered by a group of his superiors. He was designated a chief operator and shift supervisor, the job Byrnes wanted. The promotion added a few more bucks to Legg's navy pay, which was less than four hundred dollars a month. And Legg, like Byrnes, knew the promotion would pay dividends when he eventually went looking for work with civilian nuclear contractors.

By December, though, Legg had joined Byrnes in falling out of favor with his superiors. They worried he might not have mastered the technical areas beyond his electrical background. Even worse, they suspected that Legg didn't have the temperament to oversee the operation of a reactor. Months earlier, when Legg first broached the subject of promotion to chief operator, he was told unofficially not to apply. The reason cited: He wasn't ready. When Legg finally appeared before the review board, composed of military, engineering, and health supervisors, as chance would have it, one of the regular board members wasn't there—a string of family deaths back home had taken him away from Idaho Falls. If he had been, the board member said later, he would not have voted to certify Legg as a supervisor or a chief operator.

Supervisors, it seemed, had discovered Legg was almost as hotheaded as Byrnes. Bishop, former chief of the maintenance section at SL-1, worked with Legg for three months in the summer and fall of 1960 and found the young man unable to control his temper. Legg habitually flew off the handle and stormed about when things didn't go the way he wanted. Legg's temper tantrums became frequent enough that an older officer assigned to the SL-1 operation finally intervened, offering Legg some "fatherly advice" about controlling his anger. According to Bishop, the pep talk seemed to help; afterward, Legg make a conscious effort to keep his anger in check. Still, Bishop found Legg and a bad attitude were never far from each other. One day, the military supervisor of the plant entered the reactor control room and found Legg with his feet resting on the instrument console.

He commanded Bishop to instruct Legg to remove his feet. It took two orders before Legg slowly and insolently dropped his feet to the floor. Legg later complained about the order and gave Bishop "backtalk" for the rest of the shift.

Sergeant Stolla later speculated that the anger was just one facet of a personality shaped by feelings of inadequacy. Stolla thought Legg was bothered by his short stature; he thought the sailor had a raging "small man complex" that compelled him, always, to prove himself. Some workers recalled that, as a new chief operator, Legg pushed his crew members—at least the ones he didn't like—to impress the bosses. Others said he came off as a know-it-all, a guy who just had to be right.

When not in a snit, though, Legg had a personality vastly different from the serious Jack Byrnes. Legg, say those who worked with him, was a smart-ass, a jokester, a guy who liked to goad his fellow workers. And while he might have been squat, he was strong. To prove it—and he needed to—Legg would often wrestle with the other guys on his shift. He also earned the reputation at the reactor as a prankster, one not always appreciated. Stolla later remembered an incident that occurred while taking over a shift from Legg. The two were in the reactor control room when a buzzer went off on an electrical control panel. It was so like a Legg prank, Stolla said, that he didn't react to the alarm. He simply told Legg to turn off the buzzer and stop joking. Stolla was less sanguine about another of Legg's pranks, one in which he turned off a fan that cooled reactor instruments. Gauges monitoring water temperature in the reactor cooling system soon shot up, showing dangerous levels of overheating. After scaring the hell out of his coworkers, Legg reached around the instrument panel and restarted the fan to restore normal temperatures. Legg thought it was funny; Stolla didn't.

Military supervisors later admitted that they'd heard stories about Legg's pranks, horseplay, and impromptu wrestling matches. There had been talk that Legg played cards with his crew members during work shifts. They knew he didn't always do the required checks of the reactor complex. They even caught wind that Legg had at least once been sleeping in his car in the parking lot when he should have been overseeing the operation of the reactor. Sometime in the first weeks of December, the reactor's military superintendent Sergeant Lewis discovered that Legg had altered the time card of a good friend to indicate he had worked at the reactor on a particular day. The scam fell apart when Legg's buddy was spotted in Idaho Falls on that same day. The superintendent had let Legg's erratic behavior slide for a couple of months, but he now decided it was time to take action. He transferred Legg's friend to another shift and told Legg that after the holidays Jack Byrnes would be filling the vacated spot on the crew. Lewis later said he didn't know there was a history of bad blood between the two. Legg did not object to Byrnes's joining the crew, perhaps in an attempt not to make his situation even worse. Meanwhile, however, another of his superiors, his immediate supervisor, was pondering further punishment.

* * *

As Christmas approached, Legg and Byrnes weren't the only concern keeping the supervisors occupied. They were planning to shut down the reactor for the holidays, a tradition since SL-1 began operation in 1958. Though the shutdown was routine, the troubles afflicting the plant weren't. The reactor was becoming less amenable to the controlling hand of man. The mechanisms used to regulate the reactor—the five long control rods that plunged into the core—had been malfunctioning for more than a year. This was a critical problem, as the movement of the rods up or down allowed operators to control the chain reaction sparked by the more than seven pounds of uranium 235 located deep within the reactor. The bottom end of the rods contained fins made of cadmium, a metal that absorbs neutrons and slows the activity of the atoms inside the radioactive metal. As the control rods were raised, moving the cadmium, or poison, away from the enriched uranium, a nuclear reaction produced heat that turned water in the reactor to steam, which in turn was used to spin a generator to produce electricity. When operators wanted to stop the production of electricity, the control rods were dropped back into the core and the fissioning of uranium atoms was slowed to subcritical levels.

The control rods had been sticking for months, sometimes when they were being raised and often when crews tried to drop them by gravity into the reactor to stop the chain reaction of atoms. Sometimes, the rods wouldn't drop at all or would stick halfway down, requiring them to be driven down inch by inch with an electrical clutch. The plant logs revealed that in the reactor's first twenty-two months of operation, the rods malfunctioned 2.5 percent of the time—not a perfect record, but not a dire situation either. However, from November 18 to December 23, the rods stuck on drop tests 13 percent of the time. Even worse, it was hard to predict when they would glide freely and when they would seize. A report on the SL-1 operation drafted in May 1961 states that "the stickings always occurred in a very erratic and random fashion."

It was a dangerous situation, one that would have incensed the navy's Admiral Rickover and one that would have caused the immediate shutdown of his reactors. But this was the army, and it approached nuclear power in a different way. It didn't have the level of expertise or the money the navy did, and the SL-1 was in many ways just a glorified generator. It wasn't a concept that would revolutionize the army's role. Top contractors weren't lining up to eagerly bid for work. Nuclear experts weren't clamoring to get in on the project. Compared with some of the other ventures underway at the Testing Station, the SL-1 reactor was a modest proposition, an aside of the atomic age.

Supervisors of the plant, both military and civilian, believed they knew why the rods were sticking, but the answer was unsettling. They traced it back to the decision to tack-weld the thin strips of boron to the sides of the fuel assemblies that held the uranium 235. The boron, by absorbing neutrons, extended the life of the reactor core and allowed it to remain in a known, controlled state—if everything worked correctly. But investigations revealed that the delicate balance between boron and uranium was changing week to week as the boron flaked off the fuel assemblies and settled at the bottom of the reactor core. The loss of boron meant that the exact distance the control rods had to be withdrawn to start a chain reaction of the atoms changed, and crews often struggled to calculate where that subtle point was. Even more unsettling was that as the boron flaked off, the reactivity within the core increased and reduced the control rods' ability to keep the reactor core subcritical.

Operators had noticed the problem as early as 1959, less than a year after the reactor had first started operation. By August 1960, large amounts of boron were determined to be missing, and a considerable amount of boron was retrieved from the bottom of the reactor. This was not good news. But reactor supervisors worried that removing the fuel rods for a closer inspection would cause even more boron to flake off; they decided not to take any action. By November, engineers had estimated that 18 percent of the boron in the reactor core had been lost, raising the reactivity of the core. Consequently, a rod did not need to be raised as high as it once did in order to start a nuclear chain reaction.

SL-1 supervisors and the AEC, the organization responsible for monitoring the plant, had ordered piecemeal measures to temporarily deal with the sticking control rods and the loss of boron, in an attempt to keep the reactor running a few months longer. A new reactor core was expected to be installed in the spring; those in charge predicted that this would take care of both problems. In the meantime, crews were ordered to "exercise" the rods, manually lifting them up and down, in the hopes that this action would encourage easier movement when it came time for them to slide in and out of the reactor core during regular operations. To counteract the boron loss, poisonous cadmium shims were installed to reinforce the effects of whatever boron had managed to stay in place.

During the December 23 day shift, civilian engineers and supervisors helped the military crew shut down the reactor. A few of the control rods failed to drop into the reactor and had to be driven down manually. Most of the reactor operators weren't sure what to make of the situation. The equipment problems were worsening, and they weren't entirely convinced that the band-aid solutions those in charge had ordered would keep the reactor operating until the spring. Later, many of the crew members said they didn't think the problems with the reactor posed an actual danger. Unofficially, though, a few would say the reactor scared them.

The reason: The SL-1 had a unique design, one never used before or since. The reactor could be brought to a critical state, the uranium atoms in its core fissioning like a swarm of angry bees, just by withdrawing the central control rod. All other reactors required that a combination of control rods—sometimes a dozen or more—be raised in sequence to induce critical fissioning. That gave crews time to monitor the process and react if problems arose. Putting all the "worth," as nuclear engineers called it, into just one rod in the SL-1 was foolish, raising the possibility of a cataclysmic accident if the rod was moved too quickly or too far. Later, it would be recognized as a fatal engineering flaw.

Legg worked the evening shift that day, cleaning up after the day crew, making last checks, putting away tools— nothing unusual. It was just a normal shutdown, he told colleagues when he arrived at the Hotel Rogers to catch the tail end of the cadre's Christmas party. Many of the crewmen and their wives had already come and gone, and those who remained were making merry. Legg grabbed a drink and began to circulate. He ran into the reactor's training officer, Sergeant Conlon, who wasn't in the mood to wish Legg a happy New Year. Conlon pulled Legg aside and told him he wanted to see him in the administrative office at SL-1 at shift change on January 4, the next time that the two would be on duty together. Legg, his hackles rising, asked why. Conlon said he had heard about Legg's latest stunt; that letting a subordinate leave his shift and then altering the time card was the final straw. Conlon told Legg that he and other supervisors were seriously considering transferring Legg out of Idaho Falls— and maybe out of the Army Nuke program altogether. Legg became enraged. He pulled himself up to his full five feet six inches and challenged his boss to a fight. Conlon, who had broken up the fight between Legg and Byrnes at the bachelor party six months earlier, now saw the same rage turned on him. He turned away and left soon after. If Legg's fate at the reactor wasn't sealed before, it was now.

Jack Byrnes appeared to be struggling with his own demons at the party. A woman who sat at the same table as Jack and Arlene later described the nuclear worker's demeanor that night to an investigator, who paraphrased her observations in a classified memo: "He was quiet and withdrawn and his mind appeared to be on matters other than the party. She said his mind appeared to be in turmoil and he reminded her of a smoldering volcano." The woman also offered a rather peculiar observation. Byrnes had worn brown dress clothes to the party. But afterward, in her mental image of him, he was cloaked in a somber black suit.

One of Byrnes's friends from the Texaco station, Homer "Les" Clary, spent some time over the holidays with the servicemen assigned to SL-1. An army veteran himself and a former draftsman for the Boeing Company, he liked the soldiers who were coming to Idaho to work in the Lost River Desert. They were full of energy and life. They liked many of the rugged outdoor activities that the local men did: hunting, fishing, and skiing. And they knew how to have a good time. "When those guys partied, they had fun," he says. "When they partied, they partied hardy, but I don't know that they partied all the time."

But they did party over the holidays. Clary remembers in particular one holiday party thrown by the servicemen. People were drinking and having fun; it was a typical get-together. Clary chatted with Legg, whom he had not met before. Legg—a "heavy-set fellow even shorter than me"—was there with his wife, Judy, who was heavily pregnant with the child the couple was expecting in February. Clary and Legg struck up a conversation about archery. Clary had just taken up the sport, and Legg was giving him some pointers. It was a relatively tame night. Everybody was feeling good, but no one got sloppy drunk. And Legg appeared happy enough, at least to his new acquaintance.

Clary also saw Byrnes toward the end of the holidays, on New Year's Day, as he recalls. Byrnes and Clary had hit it off earlier in the year while working together at the Texaco station; their friendship jelled when each learned that the other was a die-hard skier. That winter of 1960, Byrnes had volunteered for the ski patrol at the small Pine Basin Ski Area, northwest of Idaho Falls. Clary was a ski instructor at Taylor Mountain, another small ski area southeast of town. Clary earned some money instructing, but Byrnes volunteered his time. For both men, though, the part-time jobs were essentially a way to get in more skiing without paying for it. When Jack showed up at Taylor Mountain on that first day of January, with Arlene in tow, the two men skied some runs together. The snow was good, but it was cold. A frigid blanket of air had moved into southeast Idaho with the new year, and it had yet to move east. The two men chatted mostly about skiing as they rode the chairlift to the mountain top. Byrnes seemed in a good mood. It was just another day on the slopes, Clary says.

That Sunday evening, when Jack and Arlene returned to their duplex in Idaho Falls, something happened. A television documentary released years later reported that Jack had fallen asleep on the couch and was awoken suddenly by Jackie jumping on him. Groggy and angry, he lashed out and slapped the boy. But that account has never been substantiated, and the ensuing argument could have been about money, housekeeping, or the myriad other grievances that couples use to vent their frustrations. Whatever sparked it, Jack and Arlene had a major quarrel that New Year's night. Their hapless neighbor, Robert Matlock, an engineer at another reactor at the Testing Station, had become used to the couple's fights. But this one was particularly loud, and long. Sometime before 8 P.M., Arlene knocked on Matlock's door and asked to use his telephone. Some people later speculated that Arlene called one of her husband's commanding officers. Matlock told officials later that he didn't hear the conversation. But after Arlene hung up and returned to her apartment, the angry sounds didn't resume. Matlock thinks Jack must have left the apartment while Arlene was making her call.

He was right. Jack drove to the apartment of Roger Young, a civilian engineer who was often assigned to work at SL-1. One supervisor at the reactor later said that Byrnes much preferred working with civilians than with the military folks on the crews. He must have: Young was one of the few nuclear workers at SL-1 who didn't recall ever seeing signs of a bad temper or moodiness in Byrnes. In fact, over the year just past, the two had become fast friends. They'd gone camping in the summer, hunting in the fall, and skiing in the winter. Young was also at the bachelor party where Byrnes and Legg had their confrontation, though he thought it had been a verbal, rather than a physical, dispute. When Byrnes showed up at his door that evening, angry and tired, Young took him in.

Byrnes rose early the next morning and drove to the apartment of another single friend, Martin Buckley, a soldier who was learning the intricacies of health physics at SL-1. Like so many of Jack's friendships in Idaho, this one was forged by a common love of skiing. Early on that Monday, Byrnes strapped Buckley's skis to the top of his car, and the two headed to Pine Basin. There they worked ski patrol—but mostly skied—from 10 A.M. to 4:30 P.M.

When they arrived back at Buckley's apartment, it had already been decided that Byrnes would spend the night. The two made dinner and chatted. Just before 7 P.M., Jack called Arlene. He told her his paycheck would likely be showing up in their mailbox the next morning. He wanted her to leave it in the mailbox; he'd pick it up before he went back to work at SL-1 that afternoon. Arlene apparently agreed to the plan, Buckley says. After the men ate, Byrnes went out for the evening. Buckley went to bed at 11 P.M.; Jack hadn't yet returned. The next morning, both men slept in until about 10 A.M. and then ate breakfast together in Buckley's small kitchen. Byrnes didn't say where he had gone or what he had done the night before, and Buckley didn't press him for details. But with Jack's ability to find nightlife—even on a Monday evening—it's safe to say he hadn't returned home to reconcile with Arlene.

Sometime before noon, Byrnes drove to his duplex. A

neighbor saw him check his mailbox by the sidewalk, then stand there for a moment before leaving. It wasn't long before Jack showed up again at Buckley's apartment. He was enraged and barked that Arlene must have taken his paycheck. Although he was supposed to board a bus in a few hours to put in an evening shift at the reactor, Byrnes wanted a drink. He and Buckley, who had the day off, drove to a local tavern, where they had two beers each. Buckley later said Byrnes was "emotionally upset" over his domestic problems and missing paycheck. The two left the bar at about 2:45 P.M. Byrnes told Buckley he was going to drive by his place again to see if he could get his paycheck and then take one of the AEC buses to SL-1. As it turned out, Byrnes missed the bus. He must not have retrieved his paycheck either: Earlier in the afternoon, perhaps in the bar, he'd called Kelly Calhoon, the owner of the Texaco station. Calhoon had loaned Jack eighty dollars, and Byrnes had called to say he would deposit a check that day to repay part of the debt. He didn't.

* * *

It was close to 4 P.M. on January 3 when Byrnes spun his Oldsmobile into the gravel yard surrounding the SL-1 reactor. Clothes were strewn in the backseat. He was sleeping away from home. He didn't have his paycheck, and he hadn't repaid the loan to his boss. His career was going nowhere. And with two beers in him, he was about to walk into the reactor building to take orders from Dick Legg. What a way to start a night shift. As Byrnes climbed out of the car in the waning winter light and approached SL-1, he must have

felt the cold—minus ten degrees Fahrenheit and dropping—
and heard his Oldsmobile's engine popping and ticking as it
quickly cooled down.

Byrnes and Legg might have worked together at least
briefly in the past, but no one knows for sure, and duty logs
are no longer available. Coworkers don't remember them
being paired up after the bachelor party the previous May,
and surely not since Legg's promotion in September, three
months earlier. After punching in his time card and hanging
his heavy coat in the crew quarters, Byrnes sought out Legg.
The two knew what they had to do that night. The previous
day, the day crew had gone to the second-floor reactor room.
They had moved the large, semicircular concrete blocks that
normally surrounded the reactor top to shield the workers
from any slight radiation seepage. The day crew had carefully disconnected the control rods from the motors that
raised and lowered them. Engineers had then inserted wires
into the reactor that would determine where in the core the
uranium neutrons were fissioning. It was a routine test; the
wires would be pulled out later for analysis.

The plant superintendent had issued instructions for Legg's
crew: (1) perform a reactor pump down; (2) reassemble control rods, install plugs, replace shield blocks, leave top shield
off; (3) connect rod drive motors; (4) electrically and mechanically zero rods; (5) accomplish control room and plant
startup logs; (6) perform cold rod drops; (7) check for leaks,
replace top shield; (8) perform hot rod drops; (9) accomplish
a normal startup. It was a lot of work and it required some
heavy lifting. The plant's previous supervisors, the men who
brought the reactor on-line, thought the job of reconnecting
the control rods to the drive mechanisms was potentially

dangerous. During their tenure, only they had had clearance to touch the rods. But the current supervisors at the plant had decided the work was fairly routine and could be done by any shift with a minimal crew.

Although both Legg and Byrnes had reconnected the rods before, this would be the first time Legg would do the work as a supervisor. The task was a little more interesting and challenging than much of the maintenance work at the reactor, but SL-1 managers had decided it could be done without them present. Still, reconnecting the hundred-pound control rods meant manually lifting them so they could be latched to their drives—always a touchy moment. Raising the rods, after all, was the action that sparked nuclear fission in the core. But the rods would need to be lifted only four inches to be reconnected, and that wasn't nearly high enough to start a chain reaction in the reactor.

Helping Legg and Byrnes that night was Richard McKinley. He was new at SL-1, having arrived just three weeks before from the training course in Virginia. He was only a trainee, but he was already twenty-eight years old, with a lot of military experience. He'd joined the air force right out of high school and served four years before serving with the air force reserves for a year and a half. Unlike Byrnes and Legg, he had decided a military career suited him, so he hitched up with the army. He had spent four years as a soldier, including a stint in Korea, by the time he was transferred to Idaho Falls. The bosses at SL-1 immediately liked him; he reminded them of the first wave of Army Nukes, which had included guys with some maturity and real-world experience under their belts. Legg and Byrnes didn't know McKinley well. They'd exchanged a few words and knew that McKinley was from

Ohio, had been married about five years, and was the proud dad of two young kids. About the only thing anyone later remembered about McKinley was that he was studying to convert to Catholicism. His faith wouldn't be tested the night of January 3, however—just his back. He'd be fetching tools, carrying some of the heavier items, and acting as a gofer for anything Legg and Byrnes might need.

The chief of the day shift, Sergeant Stolla, talked briefly to Byrnes before he left for the night. Stolla and his wife had once lived next to Jack and Arlene, and he was well aware that the couple had a troubled marriage. His wife, in fact, was the one who had warned Arlene that her fights with Jack were becoming too public. Stolla had also worked with Byrnes on numerous occasions and was sensitive to the younger man's moods. That night, Jack wasn't looking on his game; he appeared nervous, under strain. Stolla later said he could tell because one of Jack's eyes always twitched when he was under pressure. And it was twitching like mad as Stolla headed out the door that afternoon.

Shortly after 5 P.M., one of the Testing Station's patrolmen, Marvin Arave, pulled his car up to the chain-link fence that controlled access to SL-1. He stepped into the small guardhouse near the gate and called the reactor's control room. No one answered. He drove up to the complex and went inside the long, corrugated-metal building that was attached to the reactor silo and housed the control room. When he didn't see anybody, he walked around the site, punching the various time clocks that had been set up to ensure people were checking the plant. Arave then went back into the metal building and saw Legg and another man he didn't know. "The men seemed quite busy, and they apparently didn't want to spend time talking," Arave said later. "Sometimes they'd ask me to stop by for a cup of coffee, but not this time."

Arave would be the night crew's final visitor, though not the last person to talk to one of the crew members. Sometime that evening, probably around 7 P.M., Arlene called Jack. She later told a friend that during the phone call, the two had decided to end their marriage—Jack would never be coming back to the house. The two talked about how to split Jack's last paycheck, the missing paycheck that had so angered Byrnes just hours earlier.

Dorothy Butler was working at the Testing Station phone center that evening. All incoming calls went through her: she would call the desired phone number at the plant, then plug in a patch to connect the caller. At 8 P.M., an unidentified woman called and wanted to be connected to SL-1. No one answered when Butler tried to transfer the call. Arlene later told friend Stella Davis that she had tried to call Jack back to talk about the couple reconciling but couldn't get through. Over the next hour, Butler would take three more calls from the same woman, whom she described as increasingly frantic when no one answered at the reactor. The woman's last call came shortly before 9 P.M. When Butler again failed to reach the reactor, the response of the woman, who has never been identified in official document's, was odd: "There must be something wrong at SL-1."

Indeed there was. By 9 P.M., the three-man team had accomplished only one of the nine tasks they had been assigned. It was registered in the logbook: "Pumped reactor water to contaminated water tank until reactor water level recorder came on scale. Indicates +5 ft. Replacing plugs, thimbles, etc., to all rods."

The crew would record no other entry that night.

4

Wayward Atoms

The residents of Idaho's pastoral Snake River Valley were having an enjoyable first week of the New Year. Elder Franklin Taylor, an assistant to the Council of the Twelve Apostles of the Church of Latter-day Saints, was visiting from Salt Lake City, the Mormon mecca two hundred miles to the south. Members of the Iona Farm Bureau were gathering to view the film *Lenin's Plan for the Advancement of Communism*. The officers of the Idaho Falls Police Department were preparing to be feted at a banquet featuring a piccolo and drum duet by locals Mrs. Robert B. Harrison and her son, Jimmy.

But for the bitter cold all was right in the valley, a broad swath of potato farms and tidy Mormon towns running south to north along the eastern edge of Idaho. The Christmas and New Year's holidays had passed as ordained. Fueled by nonalcoholic punch and tracked by the watchful eyes of the elders, pious Mormon families had exhausted themselves in a whirlwind of church activities. Farm families got together with kith and kin. The men talked potatoes, the women cooked and organized, and the kids snuck out into the freezing night to make snow angels and throw snowballs.

Indeed, 1961 promised to be every bit as routine as any other year. It was the beginning of John Kennedy's Camelot; science was God; and serious government money was starting to flow into nuclear research in the Lost River Desert. As the new year dawned, life on the broad Idaho plain was placid, orderly, and prosperous.

But then, at 9:01 P.M. on January 3, an alarm sounded at the National Reactor Testing Station's main firehouse. Two long rings and then a short one interrupted the contented, post-holiday silence.

* * *

Egon Lamprecht was the youngest of seven firefighters on duty the night of January 3. On the force for just three years, he was a short, good-looking kid with an open face, a raw enthusiasm, and gift of the gab. Born into a large, devout Mormon family in Blackfoot, Lamprecht liked machinery of all kinds, especially hot rods. After graduating from high school, he opened a bicycle and motorcycle shop in his hometown. But there wasn't much of a market for two-wheeled anything in Blackfoot, and he soon got tired of eating beans all winter long.

As one of the town's volunteer firefighters, Lamprecht leaped at the chance to work at the Testing Station, even though it meant an hour's drive each way. The starting pay was a meager $3,800 a year, but the job promised security, a pension, and an escalating salary. Those benefits were hard to come by in a valley that was largely dependent on its potato harvest. And Lamprecht discovered that, except for the Testing Station's lonely location and the guards at the gates, the job was like any other in an industrial fire department. He checked fire, steam, and radiation alarms; inspected buildings; cleaned up minor chemical spills. The potentially lethal forces being unleashed in the nuclear reactors he routinely visited seemed remote.

"Unfortunately, we put the nuclear exposure thing on the back burner," says Lamprecht, reflecting on the sense of sunny optimism—and maybe complacency—that had prevailed at the Testing Station four decades earlier. "When we responded to a reactor, yeah, we took precautions. But that wasn't paramount in our mind. Nothing ever happened. If it never happens, we don't worry about it."

At the fire station, groans and mutters greeted the 9:01 P.M. alarm that indicated some kind of trouble at the SL-1 reactor, eight miles away down a dark and lonely road. The air was so frigid that night—the mercury read about minus seventeen degrees Fahrenheit—the desert crackled. The sky was pitch black. And it was the third alarm that day from the small test reactor.

Lamprecht remembers the firefighters' collective sense of dread, upon hearing the alarm, at the thought of having to head out into an unforgivingly cold desert. "Ding. Ding. Ding. Here it comes again," Lamprecht says. "Now, you're human. I'm human. It's seventeen degrees below zero. It's dark. It's not a good thing to go out at night from a warm firehouse and drive down the road. But...we do."

Six firefighters boarded a Diamond Reo fire engine and followed an assistant fire chief, who led the way in a car. They expected to find that the call was just another false alarm;

they'd traced alarms at 9 A.M. and at 2:30 P.M. earlier that day to a faulty fire detector in an auxiliary building's furnace room. Pulling into the gravel yard of the remote complex at 9:10 P.M., the firefighters saw nothing amiss—just snow, stars, and the diffused glow of lights burning in the one-story metal building attached to the glorified grain silo that contained the reactor. To their surprise, the furnace detector in the auxiliary building tested just fine.

Perplexed but not alarmed, Lamprecht, Assistant Fire Chief Walter Moshberger, and two others entered the building wearing the protective anticontamination suits and respirator masks that were a standard part of their uniform. Radiation alarms were sounding. The men had heard the alarms before, in other reactors. They knew that even small amounts of radiation could set the alarms off, and workers at the Testing Station didn't always consider the alarms to be a concern. After wiping their masks of the fog created by entering the warm building, the firefighters continued their search, working their way down a narrow corridor toward the easternmost room from which the operators controlled the reactor. Along the way they noticed lunch pails and three warm cups of coffee on the mess room table. Winter jackets were hung neatly nearby. They noticed no signs of smoke, fire, or the crew. They shouted through their masks but received no answer.

"We haven't found anybody," says Lamprecht. "This is a little spooky because there are supposed to be people on duty around the clock. We didn't know how many. But we knew someone had to be there."

Using a security radio, one of the firefighters called a nearby complex that had just been built to house another test reactor. "We thought maybe they had a problem and high-tailed it out of there," recalls Lamprecht. "When we didn't find anybody, we thought, 'Aha! They got the hell out of here and ran down there on foot.' Why else would you abandon a plant? Why would you abandon a plant that you work at unless you had a real serious problem?" But the radio check only added to the firefighters' unease: no one from the other reactor building responded to the call.

The radiation alarms echoing through the empty building suddenly took on a more ominous tone. One firefighter turned and sprinted as quickly as his bulky clothes would allow back into the night. He grabbed a radiation detector from the fire truck and rejoined his colleagues. The small box was capable of measuring gamma radiation levels of up to five hundred roentgens per hour, displaying the reading on a simple needle gauge. That level was lethal if one lingered too long in its field, and the makers of the radiation detectors hadn't envisioned the need for a machine that could read any higher levels. In 1961, it was considered safe for atomic workers to absorb a maximum of three-tenths of one roentgen per week, a standard that has been lowered over the intervening years.

Lamprecht admits that they should have been carrying the radiation detector before they walked into the building. The firefighters, their senses finally in high gear, moved cautiously into the small control room. It was empty. The instrument panel, with its large gauges, switches, and handles set into a government-gray cabinet, was a crude setup compared with today's slick, computerized control boards. The lights on the panel corresponding to the reactor's steam flow, condenser vacuum, feed-water temperature and pressure, and other critical processes were unlit, as dark as was the area beyond the command center. This meant only one thing: nothing was controlling the angry, lethal atoms in the reactor's core.

"We got [such] a reading on the detector that, by today's standards, you'd have gotten the hell out of there," Lamprecht says. "You wouldn't even get close to a field that high. You just wouldn't do it."

But in 1961, the golden age of nuclear power, nothing dreadful had ever happened at an American test site and the modus operandi was one of "No Fear." The four firefighters began climbing the enclosed metal stairs that wrapped around the outside of the windowless, cylindrical reactor building. The stairs led to the second-floor reactor room, where the top of the reactor vessel was embedded in the floor. The room contained additional metering devices and provided access to the business end of the reactor—the control rods that, when moved up or down, excited or dampened the collision of atoms.

"Starting up the stairs, we got...about halfway," Lamprecht remembers. "The radiation level accelerated so fast. The instrument reading went up just like, like a Porsche does going down the highway in a drag race, until it hit the maximum reading. With that kind of a reading, if you have any sense at all, you're going to get the hell out. We'd never seen a reading that high before."

The reading was so high, in fact, that the firefighters didn't believe it—they figured their equipment was malfunctioning. They returned to the base of the stairs, and one of the firefighters ran to get a second detector from the assistant chief's car. Armed with that, the team for a second time began to climb the cramped metal stairway. The detector pegged again at its maximum reading. "That should have told us there was nothing wrong with the instruments," Lamprecht says. "They're cool. They're telling us, 'Dummy, you don't belong here.'"

But Assistant Chief Moshberger, now ninety-seven and still living in the valley, decided to climb all the way up to the reactor floor. Accompanied by one of three health physicists who had rushed to the scene from other areas of the site, Moshberger began his ascent. Within seconds, the two had climbed into a miasma of radiation. The radiation detector that the health physicist carried also pegged at five hundred roentgens per hour. It was a level of radiation that no one at the site had ever encountered. They retreated. It was 9:35 P.M.

* * *

In the gravel lot of the SL-1 complex, the firefighters and health physicists gathered in the freezing night air. From the guard shack, they phoned their bosses, most of whom were home in Idaho Falls tucking in the kids or downing bourbon and water. Emergency call lists were hurriedly scanned. Call traffic swelled at the site's central phone dispatch. Prepared or not, the group of men in charge of SL-1 were forced to react to the crisis at hand. But any advice officials gave the firefighters for keeping the situation under control until they arrived at the site was purely theoretical. No one in the young industry had ever encountered—had ever even imagined—the scene that was now unfolding. The men gathered outside the SL-1 complex were utterly alone as they confronted the excited uranium atoms that only an hour before had been safely contained.

In the end, they were left with no choice. The crew was missing. The reactor was unguided. The building in front of them was leaking dangerous gas. Moshberger and two health physicists slipped on their respirator masks again and trundled back through the eerily quiet, deserted support building. They entered the control room and charged up the covered stairway. They didn't stop this time, even as the reading on their detector soared.

Gaining the top of the stairway, Moshberger stuck his head through the doorless entrance and observed the reactor room's operating floor. The respirator mask hampered his peripheral vision. But, as he later told Lamprecht, what he saw in that one quick scan was more than enough. The room was dim, humid, and awash in the water that should have been cooling the reactor. Small river rocks and metal punchings about the size of hockey pucks, once buried beneath the floor to shield the reactor vessel, were strewn around the circular room. On the finely crafted reactor top—a massive thing studded with bolts and housings—ports were blown open, exposing the hot core.

What Moshberger glimpsed was a deadly tableau of twisted metal. Violent forces—atomic forces—had rocked the room. And those forces were still at play. At the doorway to the reactor room, the needle on the detector was holding steady at its maximum reading. There was no sign of the crew. Moshberger turned away from a scene that would remain with him for the rest of his life.

Lamprecht still remembers the assistant chief's cryptic report to the men outside. "When he got back down he just said, 'We have a problem here.'"

Even before Moshberger ascended the stairs, Ed Vallario

VIEW OF SL-1 OPERATING FLOOR
IMMEDIATELY AFTER INCIDENT
ON JANUARY 3, 1961

The exposed reactor top following the explosion. Metal punchings
and other debris are visible in the foreground.

had received a phone call at his home in Idaho Falls, routed
through central dispatch, that warned him there was some
kind of trouble at SL-1. Vallario was the health physicist
for the experimental reactor, charged with safeguarding the
men who trained and worked there. Vallario later said he
could hear his caller getting updates from firefighters even
as they spoke, and the situation in the reactor's courtyard
was obviously hectic. Vallario felt a sick jolt of adrenaline
shoot through his body as he got the news: SL-1 heat alarms
were blaring, high radiation had been detected, and the crew
was missing. By then, duty logs had been checked, and Vallario was informed that Dick Legg, Jack Byrnes, and Richard

McKinley had been manning the reactor. Vallario knew Legg and Byrnes and had met McKinley. He liked them well enough, though he didn't interact with them socially. They were a few years younger than he and a bit less educated perhaps, but they were family men, well traveled, and bright enough to be chosen for the military reactor program. Most important, they were his charges; he was responsible for their safety. And anyone who knew Vallario knew that he was fiercely loyal.

* * *

Ed Vallario's son doesn't know what his father was doing that night, before the SL-1 reactor accident took his measure. Raised on stickball, Mozart, and urban energy, Vallario seemingly had little in common with the western landscape and the locals. Vallario had an edge, a certain street quickness in his step. The Idahoans moved through life more slowly and quietly. Brooklyn buildings kept nature at bay where Vallario grew up; Idahoans' lives were defined by the changing of seasons. Vallario played classical music on piano, an instrument he learned from his immigrant father, a professor of music; folks in Idaho Falls listened to Patsy Cline or Bob Wills, if they gave much thought to music at all.

"You know, it was kind of like 'City Kid Gone West,'" says the youngest of Vallario's two sons, Robert, who was four years old on that January night. "He was a city slicker."

But Ed Vallario also had an unquenchable curiosity about life. He later told his son that he'd found the valley's rhythms and customs interesting. "He had some friends who took him hunting," recalls Robert, "and I remember he told me he kind of got lost one day out in the hills. For several hours he followed the smoke from the fire and finally showed up in camp. The guys were pretty worried about him. In Idaho, he said, unlike some rugged country, one hill looks pretty much the same as the next. There aren't any distinguishing landmarks. Tumbleweed looks like tumbleweed, no matter what hill you're standing on."

If tumbleweed was foreign to Vallario, so too was the slide-rule view of life shared by many of the engineers and scientists he worked with in Idaho. "He was a romanticist," Robert says. "He was no nerdy engineer. He was in the midst of a community of engineers, but engineering was not his background."

After graduating from Brooklyn College, Vallario's first real job was dressing the windows of a large New York City department store. The work appealed to his sensibilities. He was an accomplished artist. He collected art—mostly commercial sketches—and scoured auction houses for antique furniture. His son remembers helping load heavy pieces of furniture into the back of Vallario's open-top Thunderbird. Searching for a career, Ed Vallario stumbled onto the then emerging field of health physics. Jobs were plentiful in the budding nuclear industry. Specialized degrees weren't required. Frontiers were being breached, rules written on the spot. It was a challenge that appealed to Vallario's quick, creative mind. He soon began working for Connecticut-based Combustion Engineering Inc. When the firm won the contract in December 1958 to take over the day-to-day operation of the SL-1 reactor from its designer, Argonne National Laboratory, the quintessential city boy decided to move west. Vallario spent the next two years setting up safety guidelines, working out emergency plans, and teaching young servicemen about the unfriendly health consequences of radiation exposure.

* * *

Minutes after Vallario received the emergency call on January 3, another health physicist, Syd Cohen, picked him up in a government Studebaker. The two then drove to the home of Paul Duckworth, the SL-1 supervisor for Combustion Engineering. Soon, the three were rocketing west across undulating Highway 20. At some point, Vallario turned to the other men and told them there weren't many options once they got to the reactor, and both Cohen and Duckworth agreed: "When we arrived at the plant site," Vallario later told investigators, "everyone was in agreement with me that we would attempt to go in there and find the men, regardless of the fact we had notice of high radiation levels."

As they approached the compound, the men found a makeshift checkpoint set up at the intersection of Highway 20—the only public highway to cut through the classified Testing Station—and Fillmore Avenue, a half-mile-long road with an oddly suburban-sounding name that headed north to the SL-1 reactor building. By then, a Class I disaster signal was being broadcast over the National Reactor Testing Station's radio networks, and buses carrying workers to the midnight shift at the various reactors on the site were being sent back to Idaho Falls. Simultaneously, top bureaucrats, generals, and admirals in Washington, DC, the men responsible for directing the country's burgeoning atomic industry, were being roused from their sleep with news that something bizarre was happening at a small, obscure reactor in Idaho.

While Moshberger and the two health physicists were in the SL-1 building, a security lieutenant, armed with a detector, noticed that radiation levels were escalating near a car by the compound's guardhouse. Those waiting for the entry team's return donned respirators, and when the men emerged from the building, all personnel relocated to the checkpoint that had been hastily set up. When Vallario and Duckworth got out of their car at the checkpoint, they were greeted with the news that the radiation detectors of the first men who had climbed the stairs to the reactor floor had "gone off-scale." Vallario's resolve didn't waver.

Vallario and Duckworth, who must already have grasped that the accident was going to devastate them professionally, grabbed high-level radiation detectors and film badges that would record their own exposures. The two men were then driven in a security car to the now deserted gravel yard encircling the reactor. As the car sped back to safety, each man donned a Scott Air-Pak respirator from the pile that had been left lying along the road. The men didn't take the time to don anticontamination suits or even gloves.

At about 10:35 P.M., Vallario and Duckworth barreled through the innocuous-looking administration building and made their way down the long hallway that led to the control room and the stairs that accessed the mangled reactor. Vallario would later remember those moments.

"As I was going up the stairs, I had the instruments in front of me, and as I got to the mid-landing of the stairs to the reactor, I heard a moaning. The plant was extremely still. There was no noise except for this moaning. I looked into the operating floor and observed Jack Byrnes near the motor-control panel...He was moaning, and he was moving. The top part of his body was twisting in an attempt to get closer to the motor-control panel. I have no idea whether or not he was aware of what he was doing, whether he was conscious. It may have been just reflex to get away from something."

The two men also saw another body—they figured it was Dick Legg—lying next to one of the concrete shields that had been moved away from the top of the reactor in preparation for the night's maintenance work. He appeared to be dead, but the two didn't have the time to verify; the needles on their radiation detectors hadn't budged from maximum levels, and they knew they'd have to come back with help to retrieve Byrnes. There was no sign of the third crew member, the trainee Richard McKinley.

"The two of us weren't sufficient to rescue Byrnes," Vallario said. "He looked like he had lost a considerable amount of blood. I immediately shouted to Duckworth, 'Let's go get some help.'"

The men ran down the stairs and out into the yard to the guardhouse, where a small group of health physicists had gathered after following them to the building. Even though they'd been warned about the high radiation levels, Vallario and Duckworth were shocked at how quickly their detectors had pegged at five hundred roentgens per hour. Vallario and other experts later estimated the true level of radiation in the center of the reactor operating room was at least one thousand roentgens per hour, perhaps even fifteen hundred, a level that would be fatal after a mere twenty minutes of exposure and that would do nasty damage to internal organs long before that.

Vallario and Duckworth's training told them that the injured man had been in the intense gamma radiation field for far too long—more than an hour and a half. Technically he was a dead man, even if he was still breathing. They also knew that they had just been exposed to the same dangerous waves of penetrating gamma and hard beta radiation that were attacking Byrnes. They hadn't worn anticontamination suits, but it hardly mattered; that kind of radiation would have passed right through the coveralls. The men's respirators were all that had kept them from breathing in alpha and beta radiation, essentially microscopic time bombs. Unlike gamma rays, alpha radiation can be blocked by a sheet of paper or even skin. Beta radiation can cause external burns but doesn't have great penetrating power. But inhaled into the lungs, alpha and beta particles are deadly, wreaking havoc on cells, bone, and blood.

All things considered, entering SL-1 again wasn't wise. Vallario and Duckworth might have made quick calculations in their heads—exposure is all about time and distance—and decided that a minute or two more in the reactor wouldn't make much of a difference. But in reality, there was no choice to make: they'd made a pact in the Studebaker as they rushed across the desert. One of the young crewmen, a man they were responsible for, was still breathing. Standing in subzero temperatures, Vallario and Duckworth knew they had to go back in. Within five minutes, they had gathered three volunteers from the group of health physicists, ran through the administration building grabbing a stowed stretcher along the way, and climbed the enclosed stairway. They did it blindly, Vallario remembered.

"We had extreme difficulty with the Scott Air-Pack units. Visibility was almost zero. All of us were more or less groping our way up the stairs. The temperature difference between outside and inside was so great that it fogged up the masks completely."

As Vallario stepped through the doorway of the reactor room and moved toward the bleeding man closest to the door, he could see that the young soldier was still alive despite the toxic environment and a horrific head wound.

"He was moaning," Vallario said. "He was not flat on the floor but seemed somewhat raised on one side with his arms, just a heaving, moaning sort of thing. If I recollect, his face was looking toward the ceiling—an entirely different position than he was in initially."

As the men hefted the crewman and carried him to the waiting stretcher, Duckworth's respirator stopped working. Fighting back a wave of panic, he made a frantic dash down the enclosed stairway, desperately trying to outdistance the alpha and beta radiation. But his lungs ran out of air, forcing him to lift his mask and inhale. With that breath, he invited a host of poison into his body; it would be years before he'd know what kind of damage it would do.

Another rescuer departed the scene soon after Duckworth to summon an ambulance. Vallario, after dashing across the twisted reactor top to confirm the second crewman was indeed dead, skittered over the debris and joined the remaining two rescuers. The three men lifted the stretcher and began to move toward the opposite side of the circular room, where an open stairway offered the promise of an easier exit. But halfway across the room, on the hottest spot on the reactor room floor, Vallario's respirator also stopped functioning.

The health physicist understood immediately the consequences. He was standing in an effluvium of gamma, alpha, and beta radiation, radioactive gas, and a dozen other radioactive by-products of the fission process. He was in nuclear hell, where enriched uranium atoms—each and every one of them an assassin—ran rampant.

Vallario's son Robert recalls his father's account of what happened next. "He said, 'I went to take a breath and it was like there was no more oxygen in the tank. There was absolutely nothing to do but take the mask off and take a breath of everything that was in the reactor containment. My best guess was to take one large breath and not to keep breathing. I basically exhaled, took the mask off, took a horrendous breath, and I did the best I could to get the hell out of there on one breath.'"

With the help of his two colleagues, Vallario, his lungs straining, hoisted the stretcher and moved quickly toward the far emergency exit. Upon reaching it, the three found the door had been blocked on the other side by machinery, preventing them from maneuvering the stretcher down the stairs. They turned, stumbled over the debris scattered across the floor, and with desperation overtaking them, bolted back to the enclosed stairway. Vallario felt as if his chest was going to burst as he helped navigate the stretcher down the stairs.

Breaking out of the stairway and into the control room, which was also contaminated, Vallario exhaled explosively and then drew in a ragged breath. He knew, though, that he couldn't expel the alpha and beta particles he'd just taken in. And then there were the penetrating gamma rays—the rays that can be stopped only by lead—that had bombarded his skin during his two forays onto the reactor operating floor.

Vallario later told investigators he had done some mental math on radiation levels in the reactor and the time it would take to get Byrnes out. He understood the risk, he said matter-of-factly, and he felt it was worth taking. The thought of abandoning the crewman never once crossed his mind. Said Vallario, "I felt he was alive, and it was necessary to go in and get him. This was essentially the thing to do."

"He often compared it to being like a fireman going into a burning building," says Vallario's fourth wife, Bette, herself a health physicist. "He felt like it was his job, his responsibility. He was focusing on the men inside." Vallario's son, a bit more crudely but with no less pride, says of his dad: "He always had big balls."

Once Vallario and the other rescuers broke out into the frigid air, they hurried the stretcher to a Dodge panel truck. One of the rescuers hopped into the driver's seat and popped the clutch, spraying gravel in the yard. A few hundred yards down Fillmore Avenue, the truck was met by a Pontiac ambulance that had come up the road from nearby Highway 20. The stretcher was transferred to the back of the ambulance and an on-call site nurse climbed in beside Byrnes. Helen Leisen, clad in a traditional white nurse's dress and shoes, leaned close to the soldier. She thought she heard a breath, ragged and shallow. She conceded later it may have been the soldier's last one. As the ambulance made a U-turn and approached the highway, Leisen slipped a respirator over Byrnes's mouth. He looked small, frail, and pale, soaked from the water that had sprayed from the reactor core. The right side of his face was destroyed.

As the ambulance sped down Fillmore Avenue, C. Wayne Bills, the deputy director of health and safety at the Testing Station, was just arriving at the checkpoint. The nuclear veteran and father of three was in choir practice that Tuesday night when a telephone call came from the site's command center. Bills immediately left the church and picked up an on-call AEC doctor, John Spickard, before heading east on Highway 20, flogging his ungainly government-issued Studebaker to its topmost speed. As the ambulance carrying Byrnes approached the checkpoint, Bills, who had just arrived, got out of his car and held up one hand to stop it. His other hand held a high-level radiation detector.

"I stopped the ambulance from going out onto US 20," he remembers. "I threw open the door. What I saw when I opened the door was just a bunch of matted, wet hair. I had a five hundred R meter, and I was reading four hundred R, and the nurse was in the ambulance. I got her out."

Spickard quickly hopped into the back of the ambulance, checked Byrnes for a pulse and a heartbeat, and then scrambled out. He shook his head. Byrnes was pronounced dead at 11:14 P.M. And Bills had a problem: The broken body in the back of the ambulance was giving off lethal amounts of radiation. It couldn't be driven to the morgue or a mortuary in Idaho Falls. Besides, the driver would absorb a fatal dose of radiation before he could get anywhere near the town. The site's medical clinic was miles away, and its decontamination equipment seemed laughable—a shower constructed under some stairs, built on the quaint notion that contaminated workers could come from the reactors and wash themselves off. Bills had to improvise.

"We got a lead blanket around the driver and had him drive the ambulance back out in the sagebrush maybe a half-mile."

Bills watched the bullet-shaped taillights of the ambulance buck and weave as it bumped through granular snow, sagebrush, and wind-scoured sand. The taillights suddenly brightened, then went dark. Bills saw the driver bail out of the ambulance and scramble into the desert to put distance between himself and his radioactive cargo. The scene was nothing short of surreal. Bills then turned and ordered health physicists to begin decontaminating the handful of workers who had made forays into SL-1. Two decontamination trailers had by that time arrived at the checkpoint. Workers were told to strip off their radioactive clothing in one trailer and then move to the other, where radiation detectors were passed over their skin and preliminary scrubbing began. Most of the initial rescue crew was sent on to the site's chemical processing plant, where they underwent secondary decontamination. Although they showered in earnest several times, readings on their hands, between their wrists and fingers especially, remained high. As a last resort, the team was instructed to scrub their extremities with potassium permanganate and Turco Decontamination Hand Soap. The purple salt and the detergent lowered the radiation readings only slightly and left their skin with a mauve tinge.

At about the same time, in the SL-1 yard, four more men—health physicists and soldiers—shrugged on respirators and masks for another foray into the reactor. Their main objective was to locate the third member of the crew. Vallario had already located the second crewman, presumably Legg, and judged him dead. But where was McKinley? Those gathered outside the reactor were perplexed. Why hadn't anyone who'd entered the shattered room seen him? Although the circular operating floor was dim, shrouded in a brown fog of radioactive gas, it was small, just thirty-eight feet in diameter, and the few lights that hadn't been smashed in the blast were still burning.

Within two minutes of entering the reactor silo, the four men rushed back down the covered stairway entrance. Their colleagues noticed that the men looked ashen and shaken. They'd found the third man, they said, after one of their flashlights lit briefly on the ceiling above the reactor. He was pinned to the metal roof above the reactor by a control rod and a heavy, stainless steel shield plug that normally kept control rods deep within the reactor core. The rod and plug, apparently driven out of the reactor with tremendous force and speed, had slammed into the crewman's pelvis and carried him into a violent collision with a steel beam before pinning him about thirteen feet above the demolished reactor top. One witness recalls a rescuer saying, "I looked up and thought I was looking at a bundle of rags hanging down."

The discovery of the third crewman put a halt to the frantic entries into the reactor. Everyone was again ordered away from the SL-1 compound and back to the checkpoint. The three men were dead, and there was no point in risking anyone else's life. As the nuclear workers gathered in small groups at the junction of Highway 20 and Fillmore Avenue, shivering, firing up Camel and Lucky Strike cigarettes, seeking warmth in the AEC Studebakers, the implications of the events they had just witnessed began to hit them. Something unprecedented had happened up there on the small hill to the north, something none of them had ever contemplated. The atoms they had corralled out there in the desert had broken free. On the loose, they had shredded building materials and wrecked the bodies of three men. And they continued to leak through cracks and under doorways and gnaw through anything not shielded by lead. It was unthinkable.

At 11:52 P.M., the site's command center took another

call. It was from a crewman's wife, asking why she couldn't reach her husband on the phone at SL-1. The woman's name has been blacked out in the log but not the message given her: "She was told no info was available at this time."

About an hour later, Stella Davis was awakened by a call from Captain James Westermeier, the commander of the military cadre at SL-1. He was top brass; Stella's husband, a sergeant, reported to him. Groggy, Stella thought the captain was calling for her husband. It took a second for it to register that he was talking to her.

"The captain ordered me to be over at Arlene's house. I think it was about one o'clock in the morning. He said, 'I want you there, and I want you there right away.' I didn't know why. It didn't dawn on me what was happening. My husband didn't know. And so I went. When I got there, a sergeant and the captain came and escorted me to the door. And that's when the captain said, 'You have to tell her.' And I said, 'Tell her what?' He said, 'Her husband has been killed.' You can imagine. It's one o'clock in the morning. It's your best friend. And having to do that is a big job.

"The captain knocked on the door," Davis recalls. "When Arlene opened it, she saw me standing there. I didn't even have to say anything. She just knew something had happened. And she just fell into my arms. They didn't really tell her what was going on, because it was top secret. All I did was keep her comfortable."

After the captain and sergeant had left, sometime close to 2 A.M., Davis told Arlene, "'Pack your clothes and get Jackie.' She stayed at my house until the plane took her back to Rome, New York. As a matter of fact, I had a birthday party a few days later, while we were waiting. It made her feel good that little kids came to play with her son, that he was able to have a third birthday party."

That early morning, similar scenes played out at the homes of Dick Legg and Richard McKinley. The trainee and his wife hadn't even finished unpacking from the move to Idaho Falls a month earlier; when Carolyn McKinley opened her door, the group of officials standing on her porch could see the boxes still stacked in neat piles over her shoulder. The military officers, cadre wives, and priests who went to the homes of the victims didn't discuss specifics. The women were simply told that there'd been an accident and that the men were still in the SL-1 building. They were kept in the dark about the gruesome scene inside the reactor silo. In the coming weeks, army and navy officers would jump through linguistic hoops to make sure the women didn't learn which man had been impaled on the ceiling.

* * *

Out in the Lost River Desert, meanwhile, officials of the Testing Station were confronted with a situation not covered in any government manual. They now had to deal with three highly radioactive bodies: one hung suspended above a potentially unstable reactor, one stretched out next to a concrete reactor shield, and one contaminating the site's new ambulance. What should be done? No one was quite sure.

Dr. George Voelz, the director of medical services at the site, recalls the extreme danger and the difficult problem the bodies posed: "All of these individuals had serious physical damage and severed limbs or worse. Each...had been struck by pieces coming off very hot reactor fuel rods." In other words, the bodies were practically glowing.

Voelz said the radiation levels in the reactor were simply too high to think about bringing the other two men out that night. Instead, the gathered officials decided to see what they could do about the body of Byrnes.

"Because his radiation level was so high, we had no place to put him," Voelz recalls. "We left him in the ambulance until we could figure out what we were going to do next. The ambulance was sitting out in the desert, amongst the sagebrush. We decided the first thing we would do is see how much radioactive material we could remove by taking off the clothing. We wanted to get this done before the morning traffic came on the site. It was about four o'clock in the morning when we decided to try to take the clothing off under the lights of a couple automobiles. Because the radiation levels were five hundred [roentgens] per hour, we decided to work outside at twenty below zero in anti-C [anticontamination] coveralls. Another fellow and I...were the first crew to try to get the [victim's] coveralls off. The legs on the coveralls had been blown up to his thighs and were all wrapped up into one big ball around his thighs; with the moisture and the cold temperatures, they were just one solid chunk of ice, having sat out there most of the night.

"The health physicists had given us about a minute's working time," Voelz says. "But we had anticipated this and had some pretty heavy-duty tools that we could use on these coveralls. We thought we could do it with two crews. Actually, we were able to get the job done with one—this other fellow and me. They had a stopwatch on us...I remember we went a few seconds over. I think it was a minute and seventeen seconds. That gives you an idea of how you have to improvise when you get into accident scenarios. We were able to get the clothing off, and we put him back in the ambulance. By that time we'd arranged where we were going to do the follow-on work. Unfortunately, taking the clothes off didn't really help anything very much. [The contamination] was about the same level as when we started."

By about 6:30 A.M. on January 4, the crew had managed to wrap the body of the first man removed from SL-1 in blankets and drape it with lead aprons. The dead soldier was to be taken to the site's chemical processing plant, where workers normally separated highly enriched uranium from spent fuel rods so the radioactive trigger for nuclear fission could be used again. The building was designed to house highly radioactive materials, although no one had anticipated that one day that would include bodies.

Firefighter Egon Lamprecht describes the strange trip the body took. "When they [health officials] decided to transport the body in the ambulance, they took readings from the driver to the victim, who were separated by a glass divider. Then they calculated the exposure to the driver based on the miles he had to drive. They decided they had to put him in a lead-lined vest. One of the firemen was going to drive the ambulance and they told him, 'You will travel X number of miles per hour,' something like seventy-five or eighty-five. Then they stopped all traffic on the highway and within the site. They gave him a free shot. And away they went."

Minutes later, the ambulance drew up at the chemical plant's large receiving bay. The driver hopped out and hotfooted it out of the area. A team of workers pulled the radioactive body out of the back of the ambulance and rushed it to a steel tank. Voelz ordered the body covered with ice and alcohol, both to preserve it and in the hopes of reducing its radioactivity, which remained distressingly high.

Throughout the day of January 4, six military men, all volunteers, rehearsed at the checkpoint a plan to retrieve the body of the second crewman. Two government photographers had been sent into the SL-1 reactor room, each for ten seconds. The pictures they snapped showed the recovery team the exact position of Legg's badly mutilated body next to one of the concrete shielding blocks that normally encircled the reactor top. Rust-colored metal punchings and rocks used to shield the reactor core lay on and around the body, and the radiation field in that area was estimated to be about seven hundred fifty roentgens per hour. Health physicists said the rescue would have to be done in stages to prevent individual radiation overdoses. They decided to use two two-man teams. Each man would wear a respirator and two pairs of anticontamination coveralls, with the wrist and ankle openings taped shut. Each of the teams participating in the retrieval of the second crewman's body would be allowed just one minute in the reactor. Two AEC health physicists would be in the adjoining reactor control room monitoring the time with stopwatches. When the minute elapsed, they would signal that the time was up with the rap of a metal pipe. A blanket was spread on the floor of the control room to wrap around the body.

At 7:30 P.M., nearly twenty-four hours after the accident, air force sergeant Max Hobson and army soldier David Arnold hurried into the reactor. Hobson later described to investigators a few of the most incredible moments of his life: "We were given one minute in time between the entrance door to the reactor floor and out again through that door; in other words, one minute on the floor. We went to the vehicle gate, through the main shop, came down the corridor to the control room. At that point, we stopped and adjusted our masks one more time, screwed them up tighter and everything.

"Then up the stairway and onto the reactor floor. We had to go around the top half of [concrete shield] block No. 1. Legg was lying on the other side of this block, between it and the reactor itself. I stepped over the body—that is, the upper portion of his body—and set him up in a seated position. At this time, Arnold took the top part of his body and I grabbed his legs and lifted him off the reactor floor. About the only thing I noticed was the shot or the punchings on the floor, red in color. We now have the body off the floor and started to leave the reactor floor."

Then, the two had a Three Stooges' moment, which, considering the grim circumstances, didn't even come close to being funny.

"A one-pound coffee can caused Arnold to trip partially near the entrance door," Hobson said. "He, by tripping over the can, caused it to move to the top landing of the stairway. At this point, I stepped in the can. Getting the body through and around the door was time consuming due to the coffee can, the body, and the air bottles on our backs. Also, during the complete trip, the nose side of my mask eyepieces were fogged over."

Hobson and Arnold manhandled the body down the narrow stairs and into the control room. They laid it on the blanket and sprinted outdoors as fast as they could manage under the weight of their bulky clothes and air tanks. Their one-minute exposure limit had expired when they were only halfway down the stairs. The second two-man team then rushed into the control room, grabbed the corners of the blanket, and brought the body outside. There, in the open air, four other men worked frantically to cut away Legg's coveralls. There was no tact involved in the process; they needed to work quickly in order to reduce their own radiation exposure and to prevent the soldier's tattered, bloody clothes from freezing in the subzero chill. The men had heard about the difficulty the others had trying to get the clothes off Byrnes earlier that morning. As with Byrnes, stripping Legg of his clothes proved fruitless: a detector passed over the body indicated that the sailor's naked torso was still throwing out radiation levels of five hundred roentgens an hour.

Bills remembers that he and another man had the last task involved in the rescue of Legg's body: "We had the door open to the ambulance and we were standing off one hundred feet or so. When they wrapped him back up in the blanket, we rushed over there, picked him up, and put him in and probably spent fifteen seconds at it."

Someone threw a lead blanket over the body, and the ambulance, with a security escort, left the SL-1 compound at high speed. It was 9:11 P.M. Once the ambulance had backed into the chemical plant's makeshift morgue, four men dragged the blanket-bound body out and rushed it to a large metal sink. Legg was slipped into an ice-and-alcohol mix alongside his crewmate Byrnes.

Now that the second body had been removed from the reactor room, health physicists, engineers, and scientists began mulling over ways to retrieve the last body from its precarious perch directly over the twisted, exposed innards of the reactor. The body hung from a steel beam, enveloped in what the experts estimated was a deadly fog of radiation exceeding one thousand roentgens per hour. Climbing onto the beam and inching out to free the body would be far too dangerous. The body itself was peppered with highly radioactive particles of enriched uranium. And there was a further complication: scientists feared that if the body or the rod holding it to the ceiling fell onto the reactor, another nuclear excursion could be unleashed.

Voelz recalls the dilemma facing recovery crews: "Until that third body was taken out, their principal concern was trying to figure out the status of the reactor. They were concerned they might do something that would cause another excursion or criticality and have a second accident. For example, whether the reactor had water in it or not was an important consideration as to whether they could have another accident. So they were trying to determine whether there was water in the reactor container itself. One of their concerns was if you had the third man—who was directly over the reactor—and you dropped him down on the reactor, would that set anything off?"

In a grotesque atomic tableau, the body of the third crewman hung from the reactor's ceiling for six days while army volunteers from a special radiological unit station in Utah figured out a way to remove it safely. They devised their plan using a full-scale mock-up of the reactor, which they constructed at a fire station training tower. On the evening of January 8, a large crane approached the exterior of the reactor and came to a halt just below a large freight door that opened onto the second-story operating floor. The crane's operator, shielded by a lead plate and guided by a spotter perched high on a pole three hundred feet away, slowly eased the big machine's long boom through the door. Attached to the boom was a five-foot by twenty-foot net. At 5 P.M., ten soldiers dashed into the reactor room to position the net under the limp remains of the serviceman hanging from the ceiling.

An official AEC photographer captured the next bizarre phase of the operation in a snapshot taken near the checkpoint at about 12:45 A.M. on January 9. The photograph shows a soldier wearing an anticontamination suit and a respirator sitting in the passenger seat of a white Studebaker. His right arm is extended out the window, and his hand grasps a long pole with a sharp hook on the end. He looks like a space-age knight preparing for a joust.

He and seven other soldiers similarly garbed and bearing long hooks were shuttled from the checkpoint to the reactor's stairway. Each one took a turn clambering up the stairway and into the reactor room. Each had sixty-five seconds to try to free the body from the ceiling, using the hooks to pull human flesh away from the metal that had impaled it. It took several attempts, but the volunteers were finally able to free the body; it dropped onto the makeshift stretcher positioned over the reactor head at 2:37 A.M.

With that ghastly task completed, supervisors decided to stop for the night; the recovery operation was suspended and the soldiers were sent for extensive decontamination. At 2:27 P.M. on January 10, the crane eased its long boom out of the reactor and slowly lowered its contaminated cargo—the third and final casualty of the SL-1 disaster—into a large cask with four-inch-thick lead lining that was waiting on a semi-trailer. The site's roads closed to traffic, security officers,

The makeshift stretcher used in the recovery of the third body.

their patrol cars flashing emergency lights, escorted the truck to the chemical plant.

One of the men assigned to help move the body from the truck recalls the transfer: "When they brought the third guy over, they brought him over in a big lead cask on the back of an International Harvester semi-truck. They gift-wrapped him in a tarp, as a matter of fact, and it was hard to get him out of the tarp because that was radioactive."

* * *

On the night of January 3, 1961, Jack Byrnes, Dick Legg, and Richard McKinley became nuclear pioneers. The three men killed on January 3, 1961, became the first humans to die in a nuclear reactor explosion. Forty-one years later, they retain the unwanted distinction of being the only humans to die in an American nuclear reactor accident.

Sickening. Shocking. Unprecedented. For the people gathered at the checkpoint at the intersection of Highway 20 and Fillmore Avenue, there didn't seem to be words adequate to describe what had happened. Nor did they suspect that the weird tale of SL-1 was just beginning.

5

"Caution: Radioactive Materials"

In the days immediately following the world's first nuclear reactor deaths, army, navy, and Testing Station officials were being asked some tough questions by politicians and the press. But perhaps the toughest question was coming from the families of the three servicemen who were killed in the reactor silo: When will the bodies be returned to us for burial? Initially, the only reply was a collection of pained and embarrassed looks from those heading up the recovery operations.

No one wanted to give the families the gory details of their loved ones' demise. Atomic death isn't pretty, and radiation doesn't die with its host. Just one look at the bodies soaking in the chemical plant's stainless steel sinks made it painfully clear that the men's remains wouldn't receive one last, lingering touch from their young wives. The alcohol and ice-water bath the men had been placed in was failing to lower the high radiation levels their bodies were emitting. Many officials suspected that the bodies were so "hot" they could never be returned to the families for a traditional burial. The chairman of the Atomic Energy Commission had even suggested the bodies be disposed of in the Testing Station's high-radiation waste pits and a small memorial be put up somewhere in the Lost River Desert to commemorate the men.

But the military's top officials at the Idaho site, reacting to growing pressure from the victims' families and with an eye toward public opinion, wanted the men returned to their families. They prevailed: traditional burials it would be—if the radiation emanating from the men's bodies could be contained. But before releasing the bodies to the families, the military wanted one last thing from its servicemen: answers. The bodies would first have to undergo autopsies.

Autopsies on bodies that had undergone such violent trauma would be, one AEC doctor warned officials, an "ugly job"—not to mention a dangerous one, considering the degree of contamination. That didn't leave the men in charge many options. In 1961, there were only a handful of pathologists in America who were even remotely experienced with radioactive bodies. Calls were made, including one by an army general, and red tape was cut. At noon on Sunday, January 8, an AEC official stood in Don Petersen's horse pasture outside Los Alamos, New Mexico. Petersen, a radiation biologist, was a top pick for the team that would dissect what remained of the last crew to work on the SL-1 reactor. The official sent to New Mexico wasted no time in convincing Petersen of the gravity and importance of the job that lay ahead. "By three o'clock I was on a DC-3 to Idaho," recalls Petersen.

Joining him were five health physicists and two other doctors, one of whom was Petersen's close friend and neighbor, Clarence Lushbaugh. All the men worked at Los Alamos Laboratory, a classified research site where America's first atomic bombs had been assembled. The eight men on board that flight had carved out careers exploring the promises and perils of the atom. They were the closest things to what could be considered "pros" in such a young and untried field. And they were about to perform three of the most bizarre and hazardous autopsies in modern medical history.

A pathologist by trade, Lushbaugh—or "Lush" as everyone called him in Los Alamos—was the official head of the autopsy team. Approaching midlife, Lushbaugh had developed a reputation as a complex, self-made man. He was a "joker," says Petersen, a man who didn't demand ubiquitous deference to his medical degree. But if you worked for him, you'd better be prepared to deliver nothing short of excellence. A female colleague later recalled Lush's wilder, more playful side, remembering a man who had a bit of a "potty mouth" and a mischievous sense of humor. Others remembered him as being extremely sure of himself, possessing a confidence that bordered on arrogance. Perhaps that air of self-assurance came from Lushbaugh's having hauled himself up by his own bootstraps. His father died in the 1919 influenza outbreak. His mother struggled financially for years afterward, and Lushbaugh relied on scholarships and part-time jobs to put himself through the prestigious University of Chicago Medical School. "He was a very active and very curious guy with a broad range of interests—essentially unbridled curiosity, very undisciplined in that regard," says Petersen.

Lushbaugh, forty-four years old when SL-1 exploded, had already made a name for himself as a "Johnny on the Spot" at radiation accidents by serving as an adviser in early, nonfatal radiation exposure cases among atomic workers. And a year earlier, Lushbaugh had participated in the autopsy of a technician at Los Alamos Labs who had clung to life for thirty-five hours after absorbing a massive amount of radiation; he had inadvertently brought together nitric acid and plutonium in a tank used to clean lab equipment, with disastrous results. But Lushbaugh seemed most proud of one of his early accomplishments, a piece of equipment he developed when working as pathologist for the Los Alamos Medical Center. One evening, as he told the story years later, he was summoned to the scene of a homicide. "A woman had been killed by her husband, who was a security policeman at Los Alamos [Laboratory]," Lushbaugh said. "He parked his revolvers in a bedside table. After a tryst in the bed one evening, he decided that she wasn't measuring up somehow or another. So, he let her have a blast in one shoulder blade that went through her heart. He became overwhelmed by his actions and he got another pistol from his bedside table and shot himself in the stomach. This bullet...penetrated his spinal cord and gave him paralysis of the lower extremities. He threw his wife across the bedroom onto a floor furnace...and called the police. They opened the door and they found all this stench and a newborn baby crawling through the blood. The open door caused the floor furnace to turn on and the woman...began to cook."

After the woman was removed from the furnace, Lushbaugh recorded her falling body temperature with a mercury thermometer but couldn't determine the exact time of death. Intrigued with the problem, he took the data he had collected to the Los Alamos Physics Lab, where technicians helped him invent the "thermistor death probe," which could be pushed far into the rectum of a body to record the central body temperature. From that reading, an exact time of death could be determined.

"Because Los Alamos was a sequestered town and had a fence around it," Lushbaugh recalled, "there were a lot of people that should not have been [there] and they subsequently killed themselves. We had a lot of murders and suicides that allowed me to use this thermistor probe as a way of finding the time of death." Lushbaugh's invention later became a regular tool for coroners.

Even as they checked into the White Horse Motel in Idaho Falls on the evening of January 8, Lushbaugh and his team knew they would have no need for the rectal probe. They knew exactly when the three men they were to autopsy had died: at 9:01 P.M. five days earlier. But they would need the kind of curiosity that had led Lushbaugh to invent the probe. They would have to positively identify the victims and determine their cause of death—had it been radiation or trauma? They would try, employing the art of forensic pathology, to reconstruct what the men had been doing and where they were doing it at the exact moment of the reactor explosion. And perhaps their most daunting task would be reducing the lethal amounts of radiation coming off the bodies. If they couldn't, there was no way the men could be returned to their families for burials.

Petersen, now in his mid-seventies and one of the last surviving members of the autopsy team, still carries the memories of the nightmarish few days it took to get the job done. The morning after their arrival, the day before the third body was removed from the reactor silo, the team was shuttled west, past frozen farm fields and into the desolation of the Lost River Desert. After going through the checkpoint near the SL-1 reactor and through the main guard station into the site itself, the men were taken into the chemical plant's decontamination bay. It was not a typical morgue, Petersen says. It was more like a big garage, seventy-five feet long and twenty-five feet high. But it had sinks that drained to contamination tanks. It also contained an overhead crane, one that could be controlled from outside the room. It would be used to move the bodies to and from the autopsy table. And there, in two large stainless steel sinks, lay two of the atomic workers, their bodies covered in crushed ice and alcohol. Shortly, they would be joined by the third victim. It had to have been a surreal sight.

On their first full day in the Lost River Desert, team members made plans for shielding themselves from horrifically high levels of radiation. There was no point in putting a Geiger counter near any of the bodies to count the ionizing particles, giving a measure of radioactivity: "It would just go wild," Petersen recalls.

Volunteers from the chemical plant helped construct a rudimentary autopsy table: a wooden plank laid across two sawhorses. Then they set about building head-high shields of lead and brick a short distance away, for the autopsy team to hide behind. "One over R squared is a marvelous thing that simply means if you can work from ten feet out, your dose is significantly lower than if you were standing right next to the body," Petersen explains. The autopsy team made hasty sketches of the tools they would need to complete their task—tools not found in a conventional medical bag or on the equipment tray in a regular morgue. The doctors would need such implements as snares, probes, shears, and cutting blades, all of which had to be attached to long steel pipes so they could be manipulated from a distance. The site's welding shop constructed the required tools "from 1½ in. wide hack saw blades, metal rod, and pipe," according to the autopsy report, in just a few hours. The finished collection resembled medieval torture devices.

On the morning of January 10, the team suited up. Their precautions hinted at the danger they faced: each man wore two pairs of anticontamination coveralls and two pairs of rubber boots, all openings taped shut. Then each donned a surgeon's cap, a full-face respirator, and two pairs of rubber gloves. Just before they entered the big bay, they slipped plastic booties over their footwear. The five health physicists were the first to enter the room, approaching the industrial sinks where the first two bodies removed from the reactor were submerged in ice and alcohol. After melting the ice with warm water and moving the bodies to an empty sink, they began to chart in an initial survey the most radioactive parts of each crewman. They then calculated how long each autopsy would take, and whether one doctor could perform it safely or whether the three physicians would need to work in shifts to avoid overexposure.

The health physicists' findings gave pause even to Lushbaugh and his assembled medical team, men who took a fairly relaxed approach to the dangers of radiation—an attitude quite typical of most nuclear workers of that era. But this was different. Flecks of uranium 235, bits of the control rods, and shards of the pressure vessel had burrowed their way into the flesh of the men, and the blast had torn their bodies into pieces.

"The Jordans [radiation detectors] were pegging at fifteen

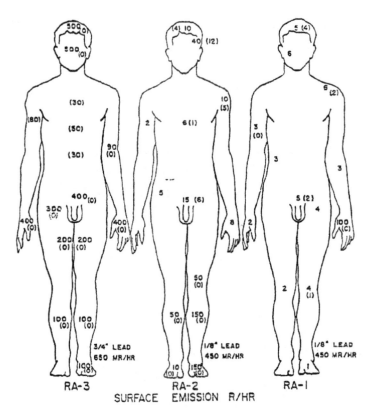

RA-3   RA-2   RA-1

SURFACE   EMISSION   R/HR

Fig. 2.   Outline diagram of the topographical radiologic
          surveys of the 3 men before autopsy. The figures
          are r/hr at the surface, and the figures in paren-
          theses are the readings after autopsy, debridement,
          and shaving. Final radioactivity after shielding
          with lead is indicated along the side of the left
          foot of each man.

–16–

OFFICIAL USE ONLY

Radiologic survey diagrams of each of the men's bodies from
the final autopsy report. The outlined figures represent,
from left to right, McKinley, Byrnes, and Legg.

hundred roentgens an hour. There was enough radiation around so there wasn't going to be a hell of a lot of monkey business," recalls Petersen.

Using long-handled snares—even broom handles at certain points—the team maneuvered a sling around the body of the first soldier recovered from the reactor. The overhead crane carried the body to the makeshift autopsy table. Petersen, who had taken medical school courses while receiving his doctorate in biology, was sent around the shield to perform preliminary procedures.

"Since I speak anatomy without an accent, I was kind of an easy guy to do the job," says Petersen. "These were preliminary preparations so they [the doctors] could go in and do exactly the critical thing, and it saved everybody's dose."

After he retreated to a safe position, one of the three physicians stepped around the barrier and made quick cuts to remove contaminated flesh and extract organs that scientists wanted to examine. In particular, samples of liver tissue could be tested for the presence and amounts of certain elements, which would help the team determine what had caused the blast and how much energy had been released.

Lushbaugh and his team made notes on the wounds found on the first body, which were substantial. A semicircular laceration at the top of the head had penetrated the complete thickness of the scalp. The lower right portion of the face had been partially destroyed. Both eyeballs were flattened and contained no fluid. The skin of the nose had been blown upward and outward, revealing the right nasal cavity. The left arm was normal, except that the left hand was shredded, with only two fingers remaining intact. The lower portion of the left leg had numerous cuts and wounds. The left side of the face, neck, and chest were a dusky violet color. The right arm and leg were free of wounds. The health physicists, after using an electric clipper to remove all the hair on the body, had determined earlier where the greatest levels of radioactivity were. The doctors set about the task of removing it.

Doctors call the surgical removal of devitalized or contaminated tissue "debridement," but this doesn't accurately describe what happened there in the chemical plant bay. The flesh that covered the entire body was highly radioactive, peppered with fission products. Great sheets of skin—some small, some extensive, some shallow, some deep—simply had to be cut away if there was to be any hope of returning the body to the family. And the doctors hadn't the time to be gentle or particularly precise. "In a nonradiation field, I suppose you could argue that it would have been done with more artistry than we were able to do hurrying," was Petersen's blunt assessment of the working conditions.

After carving away the identified patches of radioactive flesh, the team of doctors—each taking a thirty-second to one-minute turn at the body—sawed off what remained of the left hand and most of the left arm, a decision prompted by the need to further reduce the radioactivity to what was hoped would be acceptable levels. The viscera of the body—including the brain, intestines, testicles, and heart—were removed and taken to a table thirty feet away, where each organ was autopsied individually. The body was then wrapped in two layers of heavy plastic sheeting and moved to a deep freeze box. This first autopsy—which took fifteen minutes and required dozens of cuts with a long-handled hacksaw blade—would prove the easiest and most gently performed of the three.

The second body, belonging to the man found on the floor next to a concrete reactor shield, was then moved to the autopsy table. He had been far closer to the reactor than the first man when it exploded. His face had been blown inward upon impact, although there were only superficial cuts to the skin. Both eyeballs were collapsed, and the teeth were broken loose. The body's left arm was severely deformed: "The whole humerus had been shortened without breaking the skin to about half its length. The arm appeared to have been wound about this area once and then to have been unwound," cites the autopsy report. The throat was a pulpy mess, and numerous broken ribs could be felt. The upper left part of the chest was unusually prominent, while the lower part had caved in. The right knee had been dislocated and was only loosely attached to the rest of the leg by skin and ligaments. The lower half of the left leg was partially missing, and ragged shreds of skin and muscle hung from the fractured tibia. The pelvis had been fractured and was pushed up and into the body. There was severe blast damage to the lower back and left buttock. The doctors, after stripping away patches of skin studded with radioactive particles, cut out the dislocated knee, a large portion of the left thigh, and the tattered remains of the lower left leg, again in an attempt to reduce the body's level of radioactivity. Like the first, the second body was wrapped in plastic and stored in the deep freeze.

On the afternoon of January 10, the body of the third crewman was finally delivered to the chemical plant by the flatbed semi-trailer. Despite being encased in a lead-lined cask, the packaged body was spewing out massive levels of radiation. Members of the autopsy team leaned over the cask and flushed off loose bits of flesh, clothing, and reactor debris. Then they filled the box with water. The team knew the next day's autopsy on the third body would be the most dangerous, and the most gruesome. The poor kid they were going to examine had been impaled and then lifted at terrific speed into the ceiling, where he had hung over the sizzling reactor for days before anyone could figure out how to rescue him safely. There seemed to be only one way to steel themselves for the grisly job that awaited the following morning. "We were interested in heading back to the White Horse for martinis," Petersen admits.

While Petersen can't recall exact threads of the conversation that occurred over drinks that night at the hotel, it's a safe bet that they speculated about the cause of the explosion, an explosion that provided them with some of the most challenging work of their careers.

* * *

First thing the next morning, January 11, the autopsy crew drained the lead-lined cask containing the third body and again flushed the battered remains with water. The changes of water had little effect: detectors placed five inches from the head read five hundred roentgens per hour. And the health physicists estimated that if the detector were placed on the head itself, the reading would have been fifteen hundred roentgens per hour, a massive level of radiation equal to what was coming out of the exposed reactor core. There was absolutely no way the head could be buried in a public cemetery emitting that kind of radiation—the gamma rays would blast right through a casket.

The cask was drained and refilled with water and a heavy-duty detergent, and the body was soaked for another two hours. The autopsy team then attached a sling to the crane and then slipped the sling around the body. By raising and lowering the crane repeatedly, the crew agitated the body in the solution like a bundle of dirty clothing. The "wash cycle" didn't cleanse the body of much radiation. They would need to try something else.

The body was raised from the box and Polaroid photographs were taken. The photos revealed that the bulk of the head had been damaged beyond recognition by the nuclear blast. The crewman's face had simply been flattened. The nose and lower jaw had been pushed back into the neck and base of the skull. And the entire top portion of the skull, along with the scalp, had been severed cleanly—"as if by a cheese knife," the autopsy report noted—and remained attached to the head by just a thin flap of skin. The brain was exposed and severely damaged—barely recognizable as an organ that was once the center of a man's consciousness, his thoughts and emotions.

Petersen recalls the frustrating nuclear age conundrum the team faced with the third body: "Think about it. There is an object reading fifteen hundred R an hour. What do you do with it? Just being practical about it, you recognize you simply can't put that stuff out for the garbage man or put it in a medical incinerator. You couldn't even go to a crematorium and cremate it because this would contaminate the crematorium. The stuff [radiation] doesn't go anywhere. It's just there, and you have to deal with it. It would be terrible to say to the families, 'Here's the casket, but we're going to have to put it out in the middle of this field and have a funeral.' That was the consideration. What did you have to do to get the bodies' [radioactivity] down to where [the families] could not have open caskets but certainly no restrictions on the funerals?"

The way Lushbaugh saw things, he really didn't have a choice. If the body was not to be consigned to a high-level radiation waste dump, extreme action had to be taken quickly. But the only avenue left open to him was one that couldn't be traversed lightly, either ethically or politically. Lushbaugh knew that what he was about to do had the potential to be misinterpreted.

"I got a call from Dr. Lushbaugh," recalls George Voelz, then the AEC medical director at the Testing Station. "He said one of the high radiation levels was in the head and they were going to have to do some cutting and trimming to get rid of some of that. He just called me to let me know that was going to go on and asked whether I had any problem with that. I remember when he called me, I said, 'Gee, that's ugly.' But I said, 'If that's the only way we can get those levels down, you'll have to do that.'"

After having discussed over the telephone the general course of action with Voelz, Lushbaugh got back into his bulky anticontamination gear and re-entered the chemical plant bay. He and his team grabbed the custom-made autopsy tools and approached the lead box. One of the team members slipped a wire noose attached to a ten-foot-long pipe over the head of the body inside the cask. Lushbaugh had noticed that the Polaroids showed several fractures in the neck, and he positioned the remains of the head to expose those breaks. Leaning over the body with a long-handled saw—a brittle, inch-and-half-wide hacksaw blade—Lushbaugh severed the head from the body with, as the autopsy report puts it, a "rapid, sharp dissection." The head, dangling from the noose, was raised from the cask and carried to what the autopsy report calls a "lead cave," twenty feet away.

The headless body was then raised by the crane, and the team used some of the long-handled blades and hooks to rip off the shredded clothes that still clung to the body. The body was then winched to an empty sink ten feet away, where a radiation survey was conducted using high-level detectors. The team quickly realized that cutting off the head hadn't reduced the radiation nearly enough: some parts of the body were still giving off lethal levels of radiation. And the absence of clothes revealed the extent of the nuclear injuries.

The left leg had been almost completely severed from the body by a shearing force that had destroyed the left hip joint and pelvis. The right leg hadn't fared much better. Just eight inches of skin held it to the body. Ten feet of intestines protruded from a massive wound to the abdomen. The buttocks had been shredded. Both wrists were partially torn away, and the hands, colored dark blue, dangled by tendons. And the body's architecture had been completely rearranged. The upper half of the torso was twisted one hundred eighty degrees. The right shoulder was where the left should have been, and the left was where the right once was. The entire body—what was left of it—was embedded with radioactive particles.

The autopsy team found a sheet of quarter-inch-thick lead and hung it from the crane. They maneuvered the lead shield in front of the sink into which the body had been placed after the head was removed. Hiding behind the shield, Lushbaugh and an assistant studied the body closely, then ducked out from behind the lead partition to make quick cuts with an electric Stryker autopsy saw. Both the right and left legs and both hands were severed from the body. As each extremity came free, it was snared and carried, hanging from the end of a long pole, to the lead cave. What was left in the sink was then surveyed once again for radiation. The levels were no longer lethal, but they weren't low either. Yet if the crewman's family was going to get any of the remains back for burial, there simply couldn't be any more cutting. The body was rinsed a second time with detergent, then wrapped in plastic and put in the freezer with the two other bodies.

On January 12, the flesh, bones, head, limbs, and organs cut from the three servicemen were put in a fifty-five-gallon drum, which was then lowered into a lead-shielded box. The box was lifted onto the back of a semi-trailer and driven to an isolated section of the vast Testing Station grounds. The box was dumped into a deep slit trench, and a bulldozer pushed fallow desert dirt over it. An indecorous end to a substantial physical portion of the three men killed in the SL-1 explosion. The body parts had been designated high-level nuclear waste, subject to federal disposal regulations, and so joined, in sealed disposal units, the rest of the highly contaminated debris being generated by the new atomic age.

The autopsy team flew back to New Mexico on January 13 with tissue samples, notes on the locations of the three men's wounds, and a renewed respect for the power of the atom. Understandably, the families of the three crewmen were not made privy to the details of the autopsy. Nor was the public. The autopsy report, when completed, was marked "Official Use Only" and was never released to the public. But within days of the medical team's departure, rumors began to circulate among the four thousand Testing Station workers, rumors that petrified into legends in the nuclear world. The SL-1 explosion was so unprecedented, and the details surrounding it so bizarre, that gossip was only natural. Several months after the autopsies, the rumors prompted Donald Seifert of the local chapter of the Oil, Chemical and Atomic Workers Union to write a letter calling for a congressional investigation into the SL-1 incident. He took particular exception to the handling of the bodies: "Medical butchers removed glands, organs, blood and what have you for study purposes...Highly radioactive parts of the bodies were removed; heads, arms and what have you were removed and unceremoniously buried in the hot waste dump at the site."

Members of the autopsy team didn't like being called butchers. They said they had been charged with lowering the radioactivity of the bodies so the families could hold relatively normal burials. Faced with the level of lethal gamma rays the bodies were throwing out, they didn't have any other recourse but to use snares, hooks, and hacksaw blades.

Before his death in the early 1980s, army Colonel Savino Cavender, one of the three physicians on the autopsy team, tried to explain what the team faced: "Lushbaugh and I treated that individual [the third crewman] with all the respect in the world, and tried to save as much tissue as we could. We literally would have had to skin everything—all the skin off—because he had radioactive particles in so...he was so hot. It was a very difficult decision for Dr. Lushbaugh doing the autopsy, there's no doubt about it."

Petersen says that those who had not seen the bodies lying in the chemical plant's steel sinks simply could not grasp the destructive force of the nuclear blast. The third crewman, for example, had been squatting over a heavy bell housing on top of the reactor when the explosion occurred. The steel cap and parts of a control rod, followed by viciously radioactive particles, blew clear through the victim.

"The bell housing pinned him," he says. "It came up through his groin. I found one testis up in the armpit. Now that gives you an idea about the kind of trauma these guys suffered in the initial explosion."

The team left the site believing they had done a difficult job as well as could be expected. The bodies, if extensive precautions were taken, could be buried under headstones without taking anyone else with them. Furthermore, the charts detailing the blast injuries provided valuable clues about what had transpired on the night of January 3. And the team left with two surprises: one a medical curiosity, the other of more import.

The first oddity, says Petersen, involved how the bodies decomposed—or, more accurately, didn't decompose: "The thing we were worried about, particularly with the third individual, was that it was very hot up there in the ceiling, and we were afraid he was spoiling. A body left to its own devices deteriorates pretty rapidly. It turned out this was not the case. He wasn't spoiling at all. All these guys had tissue that looked like biopsy specimens. The radiation field was so high and so intense, coupled with the initial dose, the bodies were essentially radiation sterilized and would have been perfectly OK at room temperature."

The second surprise was of even more importance, and it would lead investigators down a strange path as they sought to find a reason for the explosion. During the initial survey of the accident scene, the three crewmen had been misidentified. None of them was who rescuers thought they were.

"They were going to end up burying them in the wrong place," Lushbaugh said years later.

The second body recovered from SL-1, the one lying next to the reactor, had been originally identified as Dick Legg. But the body, the doctors found, bore two tattoos. One was a red heart. The other was a flowery marking with the name "Jack" in the middle. The body was also five feet ten inches tall, a height Legg had reached only in his dreams. The body was Jack Byrnes's. The discovery meant that the first man rescued from near the reactor door, thought by Ed Vallario to be Byrnes, was either Legg or McKinley. Both men were a diminutive five feet six inches. But the first man dragged out of the reactor and taken to the chemical plant was only one hundred fifteen pounds. A tiny guy, and no one had ever called Legg that. It had to have been McKinley, the hapless trainee. That fact was confirmed when the third body was finally coaxed from the reactor ceiling and readied for autopsy. The body was stocky—one hundred sixty pounds—a weight that matched Legg's. And since Legg was the only sailor among the three, the tattoos on the third body were a dead giveaway. On the right shoulder was the letter "C," followed by the drawing of a bumblebee: SeaBee, the navy's construction battalions. The other tattoo, etched on the inside of the left bicep, was classic sailor trash, one of those ideas that must have seemed good at the time: an eight-inch-tall dancing girl.

The mix-up, in retrospect, was understandable. The rescuers were in the reactor for just a minute or two, their face masks were fogged up, and the bodies were hustled away from the reactor quickly and wrapped in blankets of wool and lead. Moreover, only a few among the rescue and recovery workers who entered the reactor had ever met any of the crew members. And the autopsy team painstakingly detailed that the bodies had been severely damaged by the blast. The men, in death, hadn't looked anything like they had in life.

The discovery of the misidentification had implications beyond just tying up loose ends—and making sure they were buried in the right plots. In the coming months, Lushbaugh and his team would use the positions in which the men had been found and the location of their injuries to reconstruct where they were and what they had been doing at the time of the explosion. The answers would add further twists to an already bizarre story.

Dr. George Voelz remembers the events that took place the day the autopsy team left: "We had a mortician from Idaho Falls come out to the site. This was after the autopsies had been done. We wanted him to come out and see what had been done and make sure that someone in the mortician line looked over what we had done. I was with him. We went out to the decontamination room at the chemical processing plant. There really wasn't any way they could embalm the bodies. He had what he called drying powder, and he suggested that in wrapping up the bodies—at least for the third individual—we just put a layer of drying powder around and then wrap the body."

The mortician had arrived at the chemical plant with three sixteen-gauge steel Batesville MonoSeal caskets, first-class commercial boxes that could be hermetically sealed. The mortician had spent days rounding up the unusual coffins after being told by site officials that it was essential they be constructed of steel and not leak air. That morning, eight volunteers donned the same kind of protective gear the autopsy team wore, pulled the remains of the servicemen from the refrigerated boxes, and began burial preparations as complex as any performed by ancient Egyptian embalmers. One by one, the corpses were wrapped in bolts of cotton spread thickly with the special drying powder, which may have dried the bodies' fluids but did nothing to preserve the flesh. The radiation had completed that task already; there was no bacteria left in the bodies to complete the cycle of life, death, and decay.

Next came layers of ubiquitous plastic sheeting. Then the real work began. The volunteers labored for ten hours to cut, bend, and form twenty-five hundred pounds of lead sheets around the bodies. The body of McKinley was wrapped in a lead sheet one-eighth of an inch thick, with extra pieces around his head to reinforce the spot where a piece of radioactive metal had slashed his skull. The "leaded package," as an AEC casualty report called the encased body, was then banded by metal strips and placed inside the special coffin in the bay with the crane. Extra strips of lead were laid in the coffin to further reduce radiation levels. Before the men closed and sealed the coffin's lid, they slipped two signs inside. One was made of cardboard and declared "Caution: High Radiation Area," and the other, made of plastic, warned "Caution: Radioactive Materials." The signs, while strange mementos for McKinley to take with him, were also chilling symbols of the macabre side of the atomic age he had unwittingly helped to usher in. The casket containing McKinley's body was then lowered and closed into a specially constructed lead vault.

The bodies of Brynes and Legg underwent the same meticulous preparations. Extra strips of lead were added to

Byrnes's coffin around the head, and the "package," already wrapped in lead and banded by steel, had another eighth-inch-thick lead sheet placed atop it before the lid was closed. The third "package," the torso of Legg, considerably smaller than the other two, was wrapped with three-quarter-inch-thick lead. Both men also went to their graves accompanied by yellow-and-black radiation warning signs.

*　*　*

The same day the autopsy team returned to New Mexico—January 13, ten days after the accident—AEC, army, and navy officials began making preparations for three very strange burials. It would take them two weeks to pull together the details; meanwhile, the caskets were stored in the decontamination room. The men certainly wouldn't be traveling to their final resting places in a commercial plane flown out of the Idaho Falls airport. The heavy vaults required special transport. And despite the layers of lead, the bodies continued to emit radiation: not much—levels were in the millirem range—but enough that officials didn't want distraught widows clinging to the boxes or priests conducting lengthy masses near them. The residual radiation also meant special precautions needed to be taken and assurances made to antsy cemetery keepers who weren't keen on the idea of nuclear oddities buried among generations of Joneses and McCanns.

By this time, Jack Byrnes's father, John, had flown to Idaho Falls to help Arlene pack up the couple's belongings. She was going back to New York with her son, and she wanted Jack buried in a family plot in a Utica cemetery. Judy Legg, eight months pregnant and devastated by the evaporation of her sense of security, was going to stay with her family in Idaho Falls. But her husband would be returning to his family in Kingston, Michigan. Caroline McKinley was planning to take her two children back home to Ohio, but she wanted her husband buried at Arlington National Cemetery in Virginia; he had served in Korea, and he had given his life while on duty. She thought him a hero and felt a burial there was justified.

On January 22, a US Air Force C-54 took off from the Idaho Falls airport with the vaults of Byrnes and McKinley tied down in its voluminous cargo hold. On board, but as far from the bodies as possible, were two army officers and an AEC official. The plane landed late that evening at Griffis Air Force Base in Rome, NY, Arlene's hometown. The vault containing Byrnes's body was removed from the plane by a forklift and put onto the back of an air force truck, which then drove to an armory garage in Utica. The next morning, the AEC official took the vault's radiation measurements. Radioactivity was detectable, especially on the bottom of the vault, but was deemed low enough for public burial. The following day, cemetery workers opened a grave with a backhoe, and a cement truck pulled up to the site. It poured concrete fourteen inches deep into the larger-than-average hole in the earth.

On the morning of January 25, the military truck bearing the vault arrived at the cemetery. The heavy box was off-loaded with a mobile crane and put on a vault hoist over the open ground. Jack's parents, brothers, sister, Arlene, Jackie, and a small group of friends gathered—at a distance—for a short Catholic burial service. After the mourners left, the vault was lowered to the concrete base poured the day before. The cement truck then revisited the grave, pouring an additional foot of concrete around each side of the vault and over its top. The next day, as soon as the concrete had set, dirt was scraped over the grave. Jack Byrnes was encased like a fly in amber. On the ground above the once-restless soldier, an AEC official stood with his detector, picking up only background radiation.

The same bizarre funeral rite awaited Richard McKinley, but since he was being buried at Arlington, it unfolded with a precision peculiar to the military. When the C-54 flew into Bolling Air Force Base near Washington, DC, at 3:30 A.M. on January 23, the aircraft taxied to a remote section of the airfield and was placed under guard. The next morning, a team of soldiers and an AEC official removed McKinley's vault with a forklift and transferred it to a flatbed truck. The vault was driven to Arlington National Cemetery—millirem radiation was detected in the cab of the truck during the thirty-minute drive—and was rolled into an unused chapel. The chapel was divided in two by a large sliding door. No one was allowed on the side with the vault, and only a guard or health physicist was permitted on the other side. During the next two nights, guards made periodic checks on the locked chapel and the AEC representative conducted radiation surveys. On January 25, the vault was driven to the grave site and positioned on a hoist above yet another hole with a concrete base. McKinley's service lasted just eight minutes, and the family was kept more than twenty feet from the vault. When the mourners departed, a cement truck rumbled through the peaceful fields of Arlington and surrounded the body of the trainee in twelve-inch-thick concrete.

Six days later, the headquarters of the military district of Washington ordered the superintendent of Arlington National Cemetery to include on the permanent record of interment

DA Form 2122 the following remark: "Victim of nuclear accident. Body is contaminated with long-life radio-active isotopes. Under no circumstances will the body be moved from this location without prior approval of the Atomic Energy Commission in consultation with this headquarters."

On the same day that the two soldiers made their journey east, a navy R5D cargo plane left Pocatello, ID, southeast of the Testing Station, with the much-diminished body of Dick Legg. After being ferried to the Tri-City Airport in Saginaw, Michigan, the lead box was trucked to a local vault company, where it was stored under lock for the night. The next morning, on January 23, the vault was driven to the Kingston Cemetery, sixty miles away. A few hours before the service, Legg's family made an unexpected request of the navy officers who had accompanied the body. They wanted to see the casket during the service. The steel coffin was lifted from its lead vault and displayed next to the open grave. Eighteen inches of concrete had already been poured into the open pit. Readings from a Juno detector showed that radiation levels doubled when the casket was lifted out of the lead. The Leggs got to see the casket. The tradeoff: The service could last only five minutes, and the family was kept at a safe distance.

Elwyn Legg, Dick's cousin and good friend, remembers that there was a bit of controversy in the small town of Kingston about the burial. The town leaders worried that Legg's body could eventually contaminate the groundwater. Some of the old-timers at the Testing Station remembered hearing that the AEC health physicists who came out of the Chicago office to help with the burial ended up gathering around the vault to convince the town it was safe to bury the remains of Dick Legg in the Kingston Cemetery.

Elwyn recalls that Dick's father, who had already lost one son, was bewildered by the death and burial of his youngest. He had been told only scant details about the explosion, and he couldn't grasp that his son and two others had been savagely killed in a novel death by a force he didn't understand. "His father told me Dick wouldn't talk about what they were doing [in Idaho], that it was a big secret," Elwyn says. "And he really didn't know much about it outside of that things could blow all to hell if something went wrong."

After the Legg family left the burial site, the sailor would suffer one final insult, cruelly subjected to the forces of physics one last time. When the cement truck arrived to pour the protective casing around the lead vault that held the casket, the operator pumped seven yards of concrete into the grave, enough to form a two-foot thick shell around Legg. But the cement had not been mixed and poured correctly; it floated the heavy vault right out of its hole.

That mishap was fodder for yet another juicy story, many of which had begun circulating back in the Lost River Desert. By the end of January, the initial shock of the SL-1 explosion had faded slightly among the Testing Station workers, giving way to incessant chatter that focused on the circumstances, causes, and aftermath of the disaster. Human nature was running its course, and consternation was beginning to give way to curiosity. This was understandable: employees worked at isolated reactors, on three shifts, for different companies and branches of the military. Few workers knew the dead; they felt no compulsion to revere their memory by quelling the gossip. For some, the technical mystery posed by the explosion was fascinating; the desert was awash with science geeks and brainiacs—how could they have been caught off guard by an event so extreme? For others, including the civilians in Idaho Falls, the salacious details that were starting to slip past the official wall of silence were intriguing. The image of the "guy pinned to the ceiling" was arresting. The initial rescue into high radiation was scary stuff, whether it was thought heroic or foolhardy. The horrific condition of the bodies, the brutal autopsies, the weird burials—it was all too provocative not to discuss.

Talk was even beginning to touch on the men's personal lives and how their private dramas may have contributed to the first nuclear disaster in American history. Only a small circle of managers knew many details about the disaster, and one former site worker says it "got awfully quiet" in the executive offices after the explosion. But that didn't stop folks from talking, from building stories and theories based solely on rumors, on the snippets of details that got bandied about as uncontrollably as the escaped atoms themselves. As Testing Station workers fanned out across the world to build a nuclear industry in the decade that followed the night of January 3, 1961, they carried the stories of the SL-1 reactor with them. Unfinished works even in life, the three young servicemen would become cardboard figures in lunchroom reenactments. Wild tales—some true and some not—would fix the crewmen in history as surely as the concrete around their bodies did in death.

In the days following the recovery of the bodies, chemical plant workers had been ordered to keep the stainless steel sinks containing the remains full of crushed ice. After numerous trips to add ice to the sinks, the workers grew accustomed to the dangerous deposits in the decontamination room. Vernon Barnes, now retired, was working in the chemical plant at the time. The story he heard was that one of the guys on the night shift slipped into the autopsy bay moments before fellow workers were scheduled to replenish ice in the tubs. Climbing into one of the sinks, the worker stretched out on the ice, lying prone just above a radioactive body. "When the guys came in, this fellow sat up real quick and straight and scared the hell out of them," Barnes says.

Barnes isn't sure if he believes the story, though he had plenty of coworkers who swore it was true. And the stories got even more graphic. According to one urban legend, Barnes says, after the autopsies were completed, a few of the employees decided to keep a souvenir: "I heard some of the guys had one of the soldier's thumbs and they were showing it around."

Disrespectful? Undoubtedly. Expected? Probably. The final bizarre twist to the SL-1 story? No.

6

Accident Aftermath

As a weak sun struggled over the Teton mountain range on the morning of January 4, the Lost River Desert looked even bleaker than usual to the nuclear workers clustered a half-mile from the SL-1 site. They knew the raft of troubles they faced were mostly beyond the ken of their collective experience. The silver silo that broke the horizon's line to the northeast was swarming with unleashed radioactive material; it would be foolish to think that none had been released into the air. They needed to know—and quickly—how much radiation had been released from the unshielded reactor; whether it was in the form of long-life microscopic particles, short-life gas, or both; and where the winds had blown the dangerous material. There was a lot of empty country surrounding the reactor, but Idaho Falls was only fifty miles to the east, Arco sixteen miles to the west, and the small hamlet of Atomic City just five miles to the southwest. Would that boosterish name hung on the dusty collection of barracks and cabins prove to be chillingly prophetic?

Just as worrisome was the question, Would the reactor provide another nasty surprise? The men coping with the aftermath of the initial explosion weren't sure that there wouldn't be a second. If any water was left in the core, it was possible it could slow the swarm of neutrons enough to cause another uncontrolled fission of whatever uranium 235 remained in the damaged reactor. How much water was belched out in the explosion? How much remained? How much would it take to excite uranium atoms? All the systems used to control the reactor were demolished. If fission began, how could it be stanched? Finding those answers depended on discovering a way to peer deep into the core without inflicting a fatal dose of radiation on another atomic worker. It was a thorny technical challenge with high stakes, for which there were no manuals and no comfortable procedures.

And, of course, the sharks were circling. The men standing in the pearly dawn were smart, and they knew anyone associated with the world's first reactor fatalities stood a good chance of being drawn and quartered professionally, either in the press, within the industry itself, or by the politicians in Washington. There was sure to be an investigation into what had caused the explosion and the deaths. Uncomfortable questions were going to be asked. Blame was going to be apportioned.

* * *

A red Cessna delivered a bit of sunshine that first bitter morning after the explosion. The single-engine plane, dubbed the "Red Baron" by some at the site and housed at the Idaho Falls airport, occasionally performed aerial surveys over the

Testing Station, including low-level runs to track radiation releases. Site managers had initially considered sending it up into the black skies shortly after the accident but decided that flight would be too dangerous. At dawn, though, the Cessna was flying grid patterns five hundred feet over the reactor, its radiation equipment sampling the air and its pilot looking for physical damage to the thin metal shell that enclosed the reactor. Almost unbelievably, the news was good on both fronts. The air-sampling gear found radiation activity no more than twice the natural background level, and the radiation appeared to be in the form of short-lived gas released through vents in the silo's roof. And the reactor's metal enclosure was intact, which meant that most of the fission materials released by the explosion were contained inside the reactor. It was amazing—and sheer dumb luck, the experts agreed. The crude cylinder, really no more than a corrugated container, was never meant to be what is called a containment vessel, the thick concrete shell that housed the larger reactors on the site and that would become a standard design feature of modern-day reactors.

C. Wayne Bills, the Testing Station's deputy director of health, had been running on adrenaline throughout the night and was exhausted by the time the sun climbed over the reactor compound. But he was buoyed by the short walk he took on the morning of January 4. Armed with a detector and protected with shoe coverings, he tramped the perimeter of the reactor silo. He picked up radiation field readings near the corrugated steel walls—high, but not excessively so—and came back from his survey with "just one little hot rock" inside the cuff of his pants. Bills concluded that the only radioactive particles to come out of the silo were those tracked out by the rescue teams. There were also some hot spots along Fillmore Avenue, small patches of radiation that had been spread by the vehicles rushing into and away from the SL-1 compound the previous night. Workers clad in anticontamination coveralls fanned out in Dodge pickups and either washed the particles into the ditches that lined the road, using high-pressure water hoses, or sealed them in resin—both crude methods by modern cleanup standards. Tests revealed that most of the particulate radiation in the compound and on the road came from strontium 90, a by-product of the fission process.

The contamination may have been limited, but radiation was released in the explosion nonetheless. Residents of Idaho Falls, though, didn't get much in the way of facts. The morning after the incident, the *Idaho Falls Post-Register* reported that "the AEC said the radiation was confined to the immediate area of the blast." But the story also contained a seemingly contradictory paragraph, one that was used unchanged in another report that ran on January 5: "Measurement of radioactivity in sagebrush samples together with highway monitoring and complex aerial surveys disclosed no contamination anywhere downwind from the SL-1 reactor."

J. Robb Brady, then editor and later publisher of the family-owned newspaper, recalls years later, "The AEC was very secretive. They wouldn't let anyone into the area to see the reactor. They just froze up about the whole thing after the first day."

Still, Brady's paper reported that the town's residents were fairly calm immediately following the accident: "There appeared to be no great concern here, 40 miles from the scene,

HEALTH PHYSICS MONITOR PERFORMING SITE SURVEY

Radiation readings are taken around the perimeter
of the reactor silo.

although the *Post-Register* and local radio stations received some calls from residents for details of the accident."

The town's almost nonchalant reaction to the accident reflected a growing ease among residents about the site's proximity and their own safety. That wasn't always the case in the early days of the Testing Station. One resident recalls that some locals were alarmed when the atomic workers started gathering at odd hours on street corners, carrying ominous black containers. Lordy, they thought, were those black boxes radiation detectors? Was the town's air being tested secretly?

It took a while, says Julie Braun, then a young girl and later a historian of the site, before residents realized the young engineers and servicemen were queuing for AEC buses—with metal lunch pails in hand.

By the time SL-1 exploded, the residents of Idaho Falls and other nearby communities were considerably more blasé about nuclear energy. Many of the locals had taken nonprofessional jobs at the Testing Station or had friends and relatives who worked there. They'd chatted with the engineers, physicists, and site managers who had moved into their neighborhoods. They seemed like decent, conservative people, and they certainly didn't have any qualms about the atom; there was no reason to believe the site was anything but a good neighbor. If the site managers now said there wasn't anything to worry about, that was good enough for most. Besides, the atomic work brought money and work and prestige into the community. The accident was going to give the Testing Station a black eye anyway; a panic attack instigated by the locals wouldn't help matters.

It would be a month before substantial details of the accident were released, but officials knew early on that radiation had indeed escaped and drifted into the Lost River Desert. In the days immediately following the explosion, the Red Baron made several air-sampling flights and roamed the lonesome countryside surrounding the site, following the route of the south-by-southwest wind that had been blowing the night of the accident. Between the plane's equipment readings and tests conducted on sagebrush, jackrabbits, the occasional sheep, and even on a cowboy and his herd of cows, officials confirmed that, despite what officials initially told the newspaper, a plume of radioactive iodine 131 gas had indeed escaped from vents in the silo and had drifted southward. A meteorological report compiled in May 1962 reasoned that the initial aerial survey failed to pick up significant radiation levels in its flight path because the weather conditions at the time—a slow-moving mass of frigid air—had held the gas close to the ground.

The news came as no surprise to the experts. As much as the industry wished it weren't so, scientists knew soon after the accident that there had been a nuclear explosion or prompt criticality in the SL-1 reactor rather than a run-of-the mill chemical reaction or equipment failure. Radiochemists had tested a small, sad pile of the crew's personal possessions recovered from the scene—a gold wedding band, a brass screw from a Zippo lighter, and the copper clasp of a watch-band. They found that the gold and copper had been activated into radioactive gold 198 and copper 64, isotopes that could be created only by a wild bombardment of neutrons.

"We hadn't faced anything like this before . . . or since," says George Voelz, the site's medical director, about the sweeping implications that suddenly confronted atomic workers at the site and the industry itself.

Almost immediately, the situation at the Testing Station became fodder for a debate then being waged in the US Congress about how close nuclear power plants should be located to metropolitan areas. The AEC feared the accident could sour the public's response to nuclear power, which leading up to the incident had been a sort of befuddled, patriotic acceptance of an esoteric science.

One day after the explosion, a headline in the *New York Times* proclaimed: "Fear Safety Question Will Produce Repercussions in Power Program." The story, noting that fifteen reactors were then in operation around the country, said the fatal incident "brought to the fore today the increasingly troublesome question of safety at future atomic power plants." Author John Finney wrote that it was the consensus of nuclear experts that plants could be operated safely in metropolitan areas. But, he reported, "Both the [AEC] and the industry have been haunted by the fear of an accident, such as the one at Idaho, that would alarm public opinion and thereby restrict the commission and discourage industry in the construction of atomic plants."

Walter Reuther, the powerful head of the United Auto Workers union, deemed it the perfect time to issue a statement saying that the SL-1 incident was one of more than forty reactor accidents that had already occurred in the nuclear industry. He warned that hundreds of civilians could have been killed by the accident if they had lived closer to the reactor. At the time, Reuther's union was before the US Supreme Court, trying to block the construction of the Fermi nuclear plant near Detroit, Michigan.

Two weeks after the explosion at SL-1, Dr. Albert Heustis, commissioner of the Michigan Department of Health, wrote to the US Public Health Service asking for "official factual data" about the explosion. Heustis complained that his department had been getting numerous inquiries about the safety of nuclear reactors. "The only information available to us so far are the press releases which in some cases have only increased apprehension and confusion," Heustis wrote. "I believe that the public health agencies, especially in areas such as ours where nuclear reactors are under construction, should receive factual data."

And then there was the inevitable tabloid-style reporting. A writer for the Scripps-Howard news service had rushed to the Testing Station and then filed a story that put a more sensational spin on the explosion: "The first authentic mystery story of our nuclear age is in the record books. Now atomic scientists are probing the great Idaho 'whatdunit' which instantly turned a tame atom into a death-dealing Frankenstein."

Industry insiders absolutely dreaded that kind of publicity. To them, the atom was no monster; it was just a building block of nature that could be harnessed and used for humans' benefit—simply another means of boiling water and spinning a generator. By 1961, they were confident in their abilities to corral the atom and keep it under strict control. There were plenty of reactors at the Testing Station—reactors a lot bigger and more complicated than SL-1—which had proven consistently reliable. A stable, safe reactor was just a matter of some complicated slide-rule work, some good engineering, and a lot of beautiful, precise welds. The explosion, though, provoked unsettling images of the atom's other face, the kind of destructive force that had leveled Nagasaki and Hiroshima just sixteen years earlier. In 1961, there was certainly none of the vociferous anti-nuke sentiment that would grow during the 1970s and 1980s. But there were nonetheless critics of nuclear power, and the SL-1 accident gave them ammunition.

On February 2, an interim report was released by the AEC that stated, "It appears a narrow plume of gaseous fission products traveled SSW from the reactor building. Low-level off-site activity of sagebrush, due to iodine 131, has been observed. Subsequent sampling in the immediate vicinity of

Highway 20 is tested for contamination
the morning after the incident.

the SL-1 facility indicates that low levels of gaseous iodine are
still being released... Close to the reactor building, soil samples
have indicated a low contamination of strontium 90."

The report proclaimed that aerial surveys had shown
radioactivity on the ground no greater than twice the background level. That statement was never contradicted, but it
didn't tell the whole story, either. In an internal memo to
the AEC's Special Operations Branch, senior health physicist
R.D. Coleman reported on other findings gathered during his
flights in the Red Baron.

First, on January 11, radioactive iodine was detected in a
suburb of Pocatello, thirty miles south of SL-1. Second, peak
readings of radiation in the sage southeast of SL-1 were not
reached until two to three weeks after the explosion. And

finally, during the second and third weeks, measurable iodine 131 was found in Idaho Falls, fifty miles west; in Butte City, twenty miles west; and in Howe, twenty miles north by northwest.

Investigators concluded that if the accident had occurred near a populated area, only those in the immediate vicinity of the plant, perhaps within five hundred feet, would have required evacuation to avoid significant doses of radiation. They conceded that if milk cows were grazing or vegetables were growing near the plant, neither the milk nor any harvest would have been fit for human consumption for several weeks because of the risk of radioiodine in the food getting lodged in the human thyroid. Still, they argued, the amount of radioactive iodine gas released had been relatively small—half the maximum permissible concentration for a 168-hour week. They said the iodine had been diluted by the air and, moreover, it had a short half-life, losing fifty percent of its radioactivity in about fourteen days as it decayed.

The findings hardly fazed Walter Reuther. He changed the tone of his initial statement, but only slightly. He opined that "thousands of people would have been overexposed to radiation if the SL-1 reactor had been built near populated areas."

A study conducted thirty years later by the US Department of Energy seemed to confirm, at least to some degree, Reuther's assertion. Scientists in 1991 reconstructed the probable path of the iodine 131 and reported that a large plume of the radioactive gas twice the background level had indeed swept through a large portion of the desert to the southeast, right to the edges of the towns of Burley and Rupert, about sixty miles away. However, the scientists also reported that the plume of iodine was one hundred times the background level at the edge of hapless Atomic City, already a developer's dying pipe dream. After coursing through the dusty town, the level dropped to fifty times background over several more miles.

But in the early days of the accident's aftermath, it was clear to both atomic proponents and critics that danger related to radiation was limited, due to the small amount of uranium used in the reactor, the plant's remote location, and the unexpected strength of the metal silo. Far more important was the explosion's demonstration of the atom's lethal force and the troubling notion that atomic workers didn't have absolute control over it. Even those within the industry were shaken by that implication.

"Our reaction was, 'How the hell did they screw it up so badly? How could you do that?'" says Clay Condit, the physicist who worked on the prestigious naval reactor program a dozen miles north of SL-1. "It came out real quickly that these guys were on top of the reactor and it blew up and a guy was pinned to the ceiling. It was kind of grisly, and it was kind of astonishing. The guys were amazed that that could happen."

Even before the second body lying in the reactor silo had been submerged in an ice bath, the AEC knew an explanation for the explosion was needed quickly. It would need answers to silence the critics, reassure the public and its own workers, and keep lawmakers firmly behind America's first high-tech industry. Accomplishing all of that would prove a challenging, perhaps impossible, task.

* * *

By 4 P.M. on January 4, the day after the SL-1 explosion, members of two commissions hastily appointed by the Atomic Energy Commission had begun flying into Idaho Falls from the East Coast. A technical advisory committee was given the task of trying to determine what had caused the accident and would report to an investigation committee that was supposed to uncover why it had happened. By January 5, members of the investigative panel were set up in a conference room at the Rogers Hotel in Idaho Falls. They began to interview everyone associated with SL-1; a long line of civilian managers, military officers, and young servicemen filed into the room, one by one, to disclose to the committee what they knew about the reactor's operations.

Some witnesses, especially those in charge of the nuclear power project, testified that there was really nothing wrong with the SL-1 reactor. Sure, they may have run up against a few technical glitches here and there. But they steadfastly maintained that any problems encountered in the day-to-day operations of the reactor were nothing out of the ordinary, nothing that wouldn't be expected in any complex industrial undertaking. They said they just couldn't understand what had happened on the night of the explosion.

Paul Duckworth, the plant superintendent for Combustion Engineering, was obviously feeling the burden of having a reactor blow up and men die under his charge. But he felt compelled, if in a flustered way, to defend his colleagues—and himself—to the committee. His emotion was evident in his testimony: "We had a good group. Something has happened. I have searched my mind many, many times to try to figure in some manner [how] myself or the group failed. We may have. I still don't know if we have—or if we have, in what way," he said.

But taken as a whole, the transcripts of all interviews conducted by the panel—interviews not available to the public until years afterward and then only through the Freedom of Information Act—show that the board was beginning to piece together another story. The investigators were finding plenty of fuel to feed criticism of the SL-1 program: technical problems that had been allowed to fester; sloppy procedures for operating the reactor; lax supervision of the trainees. There was rivalry among the three branches of the service, especially between the army and navy and between the military and the civilian contractor, Combustion Engineering, that supervised the plant's operations. Some of the lower-ranked reactor workers told the committee that they suspected promotions and choice job assignments were being handed out based on rank or service affiliation instead of appropriate qualifications, and it was damaging the morale of the military cadre.

Committee members also learned there was conflict between the military people and the civilian contractor over how much time should be devoted to training. Standard Army Nuke procedures called for eight months of academic and operational training in Virginia, then three months of academic instruction in Idaho on SL-1 operations, equipment, and health physics, followed by three months of training as a crew member under a shift supervisor. It was only after successful completion of this training and acceptable results on rigid written and oral examinations that trainees would be assigned to a shift crew. But Combustion Engineering's

Duckworth said that adherence to standard training and procedures wasn't a guarantee at the reactor.

"We have had some trouble with army people doing the work their own way...rather than the way they have been told to do it," he admitted in his testimony. Still, Duckworth told the committee that he had not been in favor of his company putting civilian supervisors on all shifts to oversee the work of the military men. He said it was difficult to "find competent CE [Combustion Engineering] people" to act as bosses. And although the AEC was supposed to be monitoring the plant's operation and management, the organization rarely got involved in operations issues. And so the problems began compounding, one upon the other. No one group seemed to be fully in charge of the SL-1 project.

"Well, there was a lot of finger pointing, and it started almost immediately," says Bills, who had participated in the recovery of the blast victims. "Most of it was aimed at the Army Reactor Branch that allowed the contractor to operate without proper supervision and essentially bypass the AEC operations office out here in terms of calling the shots. I know a branch manager for the AEC who oversaw the contract with Combustion Engineering took a real battering."

But indications of casual management and undertrained operators working on a seemingly unsafe reactor design still didn't resolve a crucial question: Exactly what happened on the night of January 3? The Washington bureaucrats, nuclear insiders, and scientists on the investigation panel wanted something specific, an equation that would yield an answer to the enigma surrounding the world's first nuclear reactor deaths.

Very quickly, the investigation began to focus on the main task the three crewmen were supposed to be doing that night: reconnecting the control rods, which essentially represented the reactor's accelerator pedal. The last entry in the reactor log revealed that the men were working on top of the reactor, preparing it for start-up. And Polaroids of the silo's interior, shot by photographers who were each allowed just thirty seconds on the operating floor in days after the explosion, indicated that the main control rod—the one that could take the reactor critical by itself—was completely withdrawn and lying on the reactor top.

In the early days of the probe, the committee learned about the well-known problems that had afflicted the rods. Its review of the site's operating logs revealed SL-1's five controls had stuck more than eighty times, and that thirty of those malfunctions occurred in the two months leading up the explosion. The logs also showed the control rods had failed to fall freely forty-six times when the reactor was "scrammed," or shut down quickly, an operation designed to mimic an emergency. The condition of the rods represented a dangerous operating environment, and it set heads wagging at more prestigious reactors at the Testing Station.

"The reactor was real tacky," says naval physicist Condit. "You had this whole miserable history of stuck rods. The naval reactor would have shut this [SL-1] down a year before. There's no way you allow that."

Army Sergeant Robert Bishop, a chief operator and head of the reactor's maintenance unit, helped shut down SL-1 on December 23 for what turned out to be the last time. Because of a series of deaths in his family, he had not been on shift work for six weeks. When he returned to work that day, he was shocked at the condition of the rods, the instrumental pieces of equipment that kept the reactor under control.

"We had had some problems previous to that with sticking rods," he testified. "But when I started to operate the plant on the twenty-third, I found the situation to be far worse than I had ever encountered before. My procedure was to push the rod drop button on a particular rod. The rod would then drop perhaps an inch or more and then stick. Then I would drive the rod in for a few inches and try it again by pushing the rod drop button. Frequently, it would drop a little more, stick again. This process continued until I had driven the rod down to about three inches, at which point most of the rods dropped readily."

Logs during that last shutdown before the holidays showed that three of the five control rods stuck and would not fall freely. They had to be driven down with their clutch mechanisms.

When asked by the committee if he had voiced his concerns, Bishop said, "I gave the opinion to CE people that during the next shutdown we should thoroughly consider whether we should continue trying to operate the plant with the rods sticking as badly as they were. I have been qualified on four reactors and feel that sticking of the control rod is a very serious problem, and something should be done about it or we should not continue to operate the reactor."

Investigators were also told about the reactor's problem with boron. Designed to "poison" or dampen the reactivity of the uranium fuel, the boron had been flaking off the metal plate lining and sinking to the bottom of the reactor. That meant the distance the rods needed to be pulled to start a nuclear reaction was changing unpredictably. It was another intolerable condition that observers said would have led to the shutdown of a reactor that was better managed.

SL-1 workers testified that supervisors, and even the AEC, which was supposedly monitoring the plant, had known about the sticking control rods and the loss of boron for months but had taken only piecemeal action to keep the reactor up and running. In fact, the official "fix" called for crewmen to lift and lower the control rods at the start of each shift until further notice. SL-1 supervisors said they thought that "exercising" the rods either with the drive mechanisms or by hand would keep the problem in check until a new reactor core was installed in the spring.

Most disturbing of all, the committee became aware that even as the control rod problems increased, the level of supervision over the men working on the deteriorating equipment was reduced. In the first year and a half of the reactor's operation, three senior men were responsible for connecting or disconnecting the control rods. In the months leading up to the accident, that delicate work had been deemed "routine" and was transferred to regular military operating crews, which sometimes consisted of just two young men with perhaps a year's experience each. Committee members also discovered that the procedures for doing that work were detailed in no more than a couple of pages of broad instructions that contained less detail and warnings than those included with a gas barbecue.

There was a feeling among some of the men assigned to SL-1 that the shutdown over the Christmas holidays had exacerbated the problems the plant had been experiencing. They suspected even more boron had been lost, raising the reactivity of the reactor further and radically lowering the number of inches a control rod needed to be raised to start a chain reaction in the core. They suspected conditions had deteriorated so drastically over the two-week shutdown that when a crewman lifted the central control rod just a few inches—either to latch it to the automatic drive mechanism, or maybe to "exercise" it—the reactor had gone critical.

Richard Feil was an air force sergeant and chief reactor operator during the winter of 1961. He had clambered onto the top of the reactor the night of January 2 to disconnect the control rods so technicians could insert wires to test the condition of the core. More than forty years later, he counts himself lucky that he wasn't the one who ended up on the silo's ceiling.

"The reactor was on the edge of stability, and it could have happened to anyone that night. It just depended on who lifted the rod," he says. "The night before the accident, I had to pull the control rods up enough to disconnect them. It could have very easily happened to me—very easily.

"We all knew that the boron was flaking off, but we didn't know how much," he says. "Of course, the engineers I guess knew how much, but we weren't privy to that information. It took quite a bit of strength to lift that center rod and latch it onto the drive, and sometimes I had to jerk to get the initial movement. But I wouldn't call it 'sticking.' It was just a matter of getting conditions right to lift the thing. The procedure was not at all unusual. I have no idea why the rod came up so far. I just know I lifted it up just enough to latch it onto the drive.

"The accident was pretty sobering to us," Feil recalls. "It was, 'There but for the grace of God go I.'"

Six days after the accident, the committee grilled army specialist Robert Meyer, a reactor operator at SL-1:

Committee member: Is there anything you feel the [investigation] board should know, either related directly to the cause, or anything you might think might be of interest to the board, realizing that the board is here to find out as much as it can about the incident?

Meyer: I'm not an engineer, but anytime you can grasp a rod and manually lift it is a poor business. That's what I felt when I first saw the arrangement they had here; this leaves it open for an incident of this sort.

Committee member: But even so, you still think most people know this so well they would be careful in doing it?

Meyer: That's right. But suppose this rod stuck a little bit; the rod is quite heavy, you know. It weighs around a hundred pounds with the shaft extension on it. Suppose it stuck a little bit and a man being slight, he probably hauled on it pretty hard. If this broke free, it's liable to come out quite a ways. This is one theory—it was sticking a little bit. We have had some trouble with the rods sticking.

Meyer's testimony pointed to one possible series of events that would have culminated in an explosion just after 9 P.M. on that January night. Meyer's responses to the committee painted a picture: One of the three servicemen assigned to the central control rod had screwed a lifting tool onto its top. Bending his knees, the crewman had gauged the effort he'd need to raise the rod. It should not have required much effort, just enough to lift the nearly one-hundred-pound rod four inches so that another worker could attach a C-clamp to it near the base of the reactor top. The crewman pulled, but the rod wouldn't rise from the reactor core. Grasping the lifting tool tighter and sinking his knees a bit more for increased leverage, the crewman applied more pressure. The rod resisted the force, then suddenly broke free, hurtling up horrifyingly fast from the reactor core. Before the crewman could release pressure, the rod had traveled far enough out of the core—already unstable and dangerously reactive from the loss of boron—to cause an instantaneous explosion.

Was this plausible? The committee members initially thought so. But as it turned out, there were a few details that couldn't be reconciled with such an accident scenario. First, the central control rod—the only rod that could have caused such devastation—was the only rod that didn't have a history of sticking. It had always slid in and out of the reactor core with ease, just as it should have. Second, the crew needed to lift the control rod only four inches to latch it to the drive mechanism. Yet the rod had to have been lifted at least sixteen inches to create the kind of prompt criticality that had occurred. Even if the rod stuck momentarily and the crewman maneuvering it applied some force, it seemed inconceivable that a fifteen-foot rod weighing nearly one hundred pounds would be withdrawn to that extreme. Third, supervisors told investigators that both Legg and Byrnes had reconnected the control rods at least four times each, and both had received extensive lecturing about never pulling a rod out too far or too quickly.

There had to be something else that would explain the inexplicable. The transcripts of interviews with SL-1 workers run for hundreds of pages, and the thought processes of the committee members often have to be gleaned by inference and tone, by the kind of questions they were asking—or by the questions they *weren't* asking—and not necessarily by the answers they were getting. Still, it's clear that the panel was probing, ever so gently, into the qualifications and personalities of the crew on duty the night of January 3. Often, committee members prefaced their questions to SL-1 workers with phrases such as, "You know, you don't have to talk to us..." or, "If you'd rather not discuss this publicly..." Not surprisingly, those interviewed by the committee often took the out being offered to them, saying they couldn't really give an opinion on the performance of Byrnes or Legg. In some cases—presumably when a crew member was willing to talk about the two—the transcripts reveal the investigation committee would then quickly go off the record, breaking off a line of questioning about the competence and temperament of the men killed in the explosion.

In the first week after the accident, the committee's interest focused on the trainee McKinley, whose body at that time was believed to be dangling precariously over the reactor. During Ed Vallario's appearance before the committee, the health physicist recounted the theory he had formed just hours after the explosion: "After understanding that McKinley was pinned to the ceiling, I felt that, knowing the operation, he had inadvertently withdrawn rod nine [the central control rod] in excess of fifteen inches and the result was an excursion."

The position of McKinley's body did seem to indicate that he was the crew member who lifted the rod. But that didn't make sense. He had been at SL-1 for only a month. He had never connected or disconnected control rods before, and other trainees were certainly not allowed to perform such a critical task. What was he doing up there on top of the reactor? The investigators began interrogating SL-1 supervisors and enlisted men about Legg, the supervising on-site operator for that evening's shift. Why would he let a trainee manipulate such a critical piece of equipment? Was he a good supervisor? Was he conscientious? How well trained was the sailor? Did he fool around—or did he let his crew do so—while on shift? The committee toyed with that possibility until January 12, when it learned from Lushbaugh's autopsy team that the bodies had originally been misidentified. As it turned out, McKinley was farthest from the reactor when it exploded, so it was unlikely he was manning the control rod. But that discovery, the committee realized, still didn't negate the theory that someone else on the crew had pulled the control rod too far, causing the explosion.

By the end of January, the investigation committee had completed its interviews, and the members returned to their regular jobs on the East Coast. They would stay in touch with each other, meet in Washington, DC, when needed, and get regular briefings from the technical committee still working in Idaho Falls. They left quietly, and they weren't inclined to share their thoughts with the locals before they went. There were rumors that the committee had brought in special investigators to poke into the lives of Jack Byrnes and Dick Legg, but no one seemed to know what the easterners were looking for or even how serious or how far the sleuthing mission was being taken. Officially, committee members said, there wouldn't be a final report until the autopsy team and the technical advisory committee could reconstruct the crew and reactor's final moments.

* * *

In a flurry of long hours punctuated by frantic, minute-long forays into high radiation fields, the first phase of post-accident operations—the initial emergency response, retrieval of the bodies, and monitoring of the radiation levels in the environment—had taken just days. The second and third phase of activity—determining the stability of the damaged reactor and pinpointing the cause of the explosion through physical evidence recovered from the scene—would take considerably longer. Combustion Engineering would remain in charge of investigating the reactor's status. But it was clear that CE's role in managing the plant would be questioned during the investigation, so once the reactor was declared stable, the General Electric (GE) Company would take over gathering evidence from the site and then demolish the rest of the reactor.

At the outset of February 1961, there was still long-life radiation zipping around inside the SL-1 reactor building, making the immediate area unsafe for the enormous group of personnel required for the final two phases of the operation. As a result, the checkpoint erected at Highway 20 and Fillmore Avenue on the night of the accident was expanded to become the official command center for the months of activity ahead.

In addition to the decontamination trailers already in use at the checkpoint, two dressing trailers were hauled to the location. In the first, crews suited up in layers of protective clothing and donned respirators. Then they moved into the next trailer, the "Buffer Zone," where their equipment was checked and individual radiation exposure badges were issued. From the second trailer, the crews emerged in the "Hot Zone" and were shuttled the one mile down the road to the SL-1 site. When their task at the reactor was completed, they returned to the checkpoint to begin decontamination procedures.

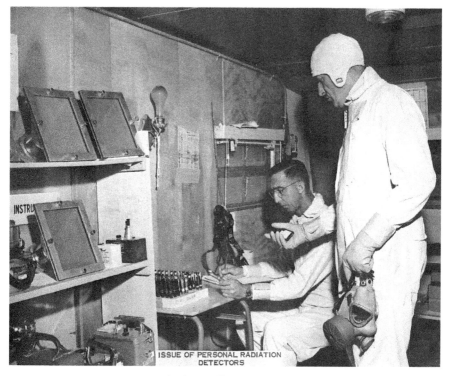

ISSUE OF PERSONAL RADIATION
DETECTORS

Inside the "Buffer Zone" trailer.

Two other trailers that were brought to the checkpoint served as administration offices for CE and GE supervisors, and yet another became a makeshift electronics maintenance shop and communications center for keeping in contact with people working at the reactor.

With the command-and-control hub up and running, the search for water inside the reactor core began. Water plays a crucial, three-pronged role in a reactor's operations. First, it carries heat away from the fireball of fissioning uranium. With the SL-1 design, the water also boils, which creates the steam that turns the generators and produces electricity. Finally, water acts as a "moderator," slowing down the dance of neutrons fired from the enriched uranium so that they slam into other atoms at the right speed to split off even more neutrons. The process continues until the core crackles into a self-sustaining nuclear reaction. If water was still trapped inside the damaged SL-1 reactor core, it was conceivable that the enriched uranium atoms left in the core could suddenly begin to fission. With the control rods completely destroyed by the explosion, there would be no way to control the reaction. But if the core was devoid of residual water, if it had all been spewed out in the violent initial blast, the reactor was likely stable.

On the night following the accident, a first crude attempt was made to determine whether there was water inside the reactor. Resorting to primeval methods, the nuclear engineers and physicists instructed a soldier who was helping recover the second crewman to toss a rock into one of the open ports in the reactor top and listen for a splash. It was yet another unbelievable moment: cutting-edge nuclear technology reduced to a carnival pitch game. The soldier, hampered by a mask and time, couldn't manage to underhand a rock into the porthole. The scientists would have to devise another way to test for water.

The team faced the perplexing problem of needing to survey the reactor without sending crews directly into the silo—the radiation levels were just too high. Using the full-scale mock-up of SL-1 that had helped crews plan the retrieval of the third crewman's body, the team practiced making remote entries into the reactor using TV cameras, drop lights, and a host of other specially adapted instruments. The equipment was attached to a twenty-five-foot traveling boom powered by a hydraulic crane. Throughout the months of February, March, and April, remote procedures were rehearsed and reviewed until every worker had perfected his role in the operation.

By late April, the crews were ready. Each group received a final briefing before being transported down Fillmore Avenue to SL-1. The crane was positioned at the back of the reactor building and the boom, decorated with what was high-tech equipment at the time, entered the silo through the second-story freight doors, guided by an operator shielded in the crane's lead-lined cab. The TV cameras, also shielded by lead plates to prevent the film from fogging up in the high gamma-ray fields, revealed what many workers had caught only glimpses of in the Polaroid photographs taken right after the explosion: a violently twisted, viciously radioactive mess.

When operators maneuvered the cameras through the open ports and into the guts of the reactor, the images beamed back seemed to show a core that was bone-dry. But the team needed to be absolutely sure. It performed two more remote penetrations of the reactor vessel. The first was with an ultrasonic vibration probe; when lowered to a depth only six inches above the top of the core, its frequency suggested that the vessel was dry. Those results were confirmed by a second probe, one covered with a water-soluble chemical. It returned from the bowels of SL-1 with its chemical component intact. There was no water left inside the reactor.

With the core deemed stable, crews quickly shifted their focus to dismantling the reactor and discovering what secrets it held. That task fell to hundreds of GE employees, who a month earlier had faced unemployment when President Kennedy canceled their work on the air force's ill-conceived airplane reactor project at the site. The cancellation of the project meant that a massive hangar equipped to handle high-

PERFORMING AN ENTRY INTO THE SL-1 REACTOR BUILDING

Performing a remote entry.

level radioactive waste was standing empty. Tearing down SL-1 would be tortuous work, but at least it was work. Because radiation levels were still so high, the time employees spent inside the reactor silo would be strictly limited, and the exposure spread out over a significant number of people.

Beginning in June, a conga line of 1,240 people—GE employees, soldiers from Utah, and other site employees looking to pick up some overtime work—volunteered to take their quarterly maximum radiation dose. Decked out in canvas and plastic coveralls; surgeons' caps, hoods, and respirators; rubber boots, shoes covers, and cotton and plastic gloves to protect themselves from beta and alpha radiation, volunteers climbed the stairs to the second floor of the silo and did what they could in the two or three minutes they were allotted. Some loosened bolts just enough for the next workers to dash in and remove them. Others made short passes over the debris-littered floor with industrial vacuums before handing the nozzle to someone else. Health physicists stood outside at the bottom of the steps—eighty of them took rotating shifts to ensure their specialized skills wouldn't be lost because of overexposure—and banged on the railing when the workers' time was up. On average, it took workers four hours to suit up and strip down—all for those few fleeting moments of cleanup inside the reactor building. GE later estimated that throughout the operation, it used and disposed of approximately nine thousand pairs of shoe coverings.

Bit by excruciating bit, the reactor floor was stripped of tools, rags, and small indistinguishable lumps that may or may not have been body parts. Pieces of the four control rod mechanisms ejected from the reactor were recovered, including the all-important central rod that had been lying

BRIEFING THE ENTRY CREW

The command center at Highway 20 and Fillmore Avenue.

across the top of the reactor. The debris was sealed in fifty-five-gallon drums and trucked miles across the desert to the giant hangar, called the TAN (Test Area North) Hot Shop. Recovered items that could potentially help solve the SL-1 puzzle were kept there; the rest were taken back to the SL-1 compound for disposal.

As larger pieces of machinery were hauled out of the building and the steel walls and ceiling were torn down, radiation levels in the yard around the site began creeping to unacceptable levels. Officials considered trucking the radioactive waste to what the Testing Station called the Burial Ground, the final resting place for all the site's really bad stuff. But it was a sixteen-mile journey, partly over public Highway 20.

Carrying dangerous cargo over such a heavily trafficked road posed too much risk. And it would mean too much exposure for the drivers. Instead, two large pits and a trench were dug among the sagebrush, about sixteen hundred feet from the SL-1 compound. Over the course of the sweltering summer and into the fall of 1961, a dump truck, its driver protected by lead sheets, hauled the remains of the reactor into the desert. The debris filled fifteen hundred feet of gouged earth. Bull-dozers kept piling dirt into the pits until a health physicist no longer picked up radiation readings with his detector.

In November, a huge crane lifted out the soul of SL-1: the engine, what nuke workers call the pressure vessel. As the core was hoisted from the decapitated silo, workers silently looked on, wondering how such a small structure—just four-teen and a half feet tall and four and a half feet in diameter—could have caused so much destruction. The freed core was slipped into a tall steel container which was waiting on the back of a flatbed truck. It was hauled slowly, with a security escort, to the Hot Shop.

C. Wayne Bills, the AEC's deputy director of health and safety at the Testing Station, had been appointed to oversee the work of the technical group in charge of probing for the cause of the SL-1 blast. He says it was the removal of the reactor core that allowed scientists and engineers to fit the final pieces of the puzzle together. It would take them almost a year, but the technical investigators eventually ruled out some of the leading theories proposed in the first weeks after the explosion. Tests showed that the reactivity of the core had not changed appreciably over the Christmas shutdown as site personnel originally thought. This finding seemed to rule out speculation that the reactor was unstable and on the verge of criticality on the night of January 3, needing only a few inches of control rod movement to explode like some gag cigar.

Bills's group also ruled out the possibility that the explosion was anything but a nuclear excursion. Despite initial radiological evidence that a nuclear event had taken place within the SL-1 silo, there were a few holdouts at the Testing Station who thought another force could have been responsible for the blast. Earlier, the AEC manager of the Idaho site wrote a memo to one of his bosses in Washington, DC, asking that the FBI be brought into the investigation. Allan Johnson wrote, "Although there is no evidence for support, we are thinking about sabotage possibilities." The memo was forwarded to the investigation committee, but the chairman couldn't get a consensus on bringing in FBI agents—with the inevitable bad publicity that would result. Instead, the committee hired private explosive experts, who worked quietly, away from the public and media's eye. T.C. Poulter of the Stanford Research Institute eventually eliminated the possibility of an explosive charge having been placed in the reactor: "There was no sabotage involved in this event...which could have been caused by a chemical type of explosion."

But after months of sifting through debris pulled from the reactor, the technical team uncovered some concrete, telling clues. One of these clues, the finely machined central control rod, had been lying in the Hot Shop since the first days of the investigation. But it took some time to crack its secrets. The top part of the rod, called the rack, was still wrapped in the metal sheath in which it traveled up and down. The rack was stuck in the sheath, with just four inches protruding: exactly the height it should have been if the

SL-1 Reactor Removal

The SL-1 reactor is removed.

crew was reconnecting it correctly to the drive mechanism. "When we found it seized in the right place, we thought, 'Gee, there must have been an explosion down in the vessel that blew the center rod out.' We were still very much in a quandary," Bills explains.

Some time later, the members of the team were looking over debris recovered from the fan room, a small area located directly above the reactor. A nondescript length of pipe caught their attention. It was the straight pipe that crews attached to the top of the rack to lift the control rods. Attached to the pipe were a nut and washer and a bit of the rack. It had obviously been blown through the ceiling by the explosion and had lain there for months. And it spoke volumes to the engineers: It was the clue they'd been looking for.

"When the handle got thrown into the attic, it was a missing piece of evidence," Bills says. "We didn't really know what part of the operation the crew was in. When we found the handle and discovered it was hooked above the washer and the nut, we knew they were in the final stages of assembly, [when] you were only supposed to raise the rod a quarter of an inch. You weren't even supposed to raise it four inches."

The discovery meant Legg's crew had, sometime before the blast, already raised the control rod—the delicate part of the operation—a bit more than four inches and installed the C-clamp to hold it temporarily. The handle was then removed and a nut and washer installed to connect the control rod to the drive motor. The evidence indicated that that work had been done and the handle reinstalled. The crew simply had to take pressure off the rod to remove the clamp. At that point, they were just a quarter-inch lift from a routine night and likely long, prosperous lives.

Bills and his team turned their attention back to the control rod, still highly radioactive. Could it be hiding the explanation for what had happened in the millisecond before the explosion? They decided to dissect it by cutting away the sheath. There, inside, on the normally smooth surface of the rod, they found scratches. Those marks would forever damn at least one member of the SL-1 crew on duty the night of January 3.

From a methodical examination of the piece of equipment, Bills's team was able to make a certain deduction: "The central control rod had been jerked out," he says. "The first scratch is when the thing is pulled out twenty inches. It was then blown up to the ceiling and hit a beam. That's where the handle broke off and the rod was driven all the way back into the sleeve. When the rod came back down and hit the reactor vessel lid, the rod was driven back up to four inches, exactly where it should have been. Until we cut the thing apart, we didn't know it had been out to twenty inches and we didn't know it had gone back to zero. It had scratches back down to zero and then scratches back up to four inches. I almost got away with saving that piece, but it was still radioactive. I couldn't talk them into saving it, but that was a real gem."

The discovery proved that one of the crew members had suddenly pulled the central control rod out twenty inches. Although there was no other plausible explanation, it seemed inconceivable to investigators. There simply was no reason to ever pull a rod out that far. And the technical team now knew the rod had already been withdrawn four inches with no problem and that it only needed to be eased back down. That seemed to disprove the theory put forward by some SL-1 cadre members that the rod must have stuck when it was first being lifted, forcing the crewman gripping the handle to pull up on it too hard.

Still, Bills's team decided to do one more test. It built a mock-up of the reactor top, complete with a dummy control rod weighing about the same as the one found in SL-1. Team members then positioned men below the mock-up to hold the control rod, mimicking the resistance a crewman would

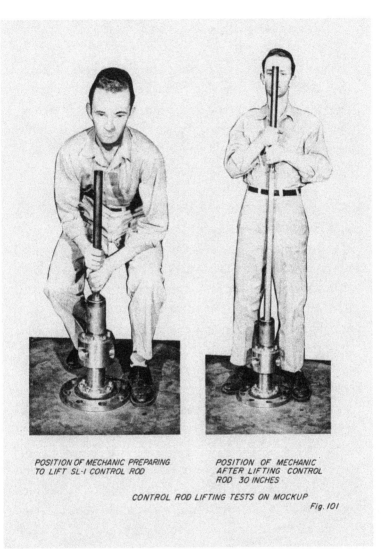

POSITION OF MECHANIC PREPARING
TO LIFT SL-I CONTROL ROD

POSITION OF MECHANIC
AFTER LIFTING CONTROL
ROD 30 INCHES

CONTROL ROD LIFTING TESTS ON MOCKUP

Fig. 101

From the official interim report.

feel if the rod had stuck. Men of all sizes and strengths were brought in and told to grab the lifting handle and pull as hard as they wanted. As each yanked up, the control rod would be suddenly released from below, as if a sticking rod had just let go. Not one of the men pulled the rod far enough before stopping its upward progress to have created a nuclear explosion.

"They would only override by maybe ten inches or so," says Bills. "No one got to the twenty-inch mark. I don't think you'd overshoot it that far. Once the rod went over fourteen inches, the reactor went prompt critical. The fact that it went to twenty inches was incidental. By the time he'd gone up to twenty inches, the reactor was going."

By that time, Bills had been tipped off about Legg's reputation as a prankster with a penchant for horseplay and wrestling. Though it seemed like a ludicrous experiment, Bills felt compelled to confirm that the world's first nuclear reactor deaths hadn't been the punch line of a joke gone terribly wrong . "We looked at goosing," he says. "You know, someone grabbing the guy on the rod in the rear and having him jump."

In what must have been yet another surreal SL-1 scene, one volunteer after another stepped onto the mock-up reactor, began pulling up on the rod, and was given a goose—a poke in the butt—delivered as unexpectedly as possible under the conditions. Not one of the men jumped or jerked or yanked hard enough on the rod to raise it anywhere near the critical fourteen inches. And good thing, too—the whole affair was shaping up badly enough as it was; it wouldn't do to discover the gruesome deaths had been the result of a prank. Still, it was only marginally good news, because for Bills, his team's finding could only mean one thing.

"It all came down to the center rod being pulled up too far by a crewman," he says. "I don't think you'd overshoot it like that. The fellow could have stopped it if it wasn't some kind of just deliberate 'to hell with it' action."

By the time Bills figured out what had happened mechanically the night of January 3, Don Petersen and the autopsy team had figured out who had done it. The team had returned to Los Alamos knowing how each man died—from multiple traumas—but they needed those horrific wounds to tell them a story: Who was where and doing what when the blast ripped the silo apart? Lushbaugh and the other doctors thought they had figured it out based on the injuries they had observed, but they looked to Petersen to confirm their hunches. During the autopsies, the biologist had been collecting hair samples from the victims: from their legs, groins, arms, and heads. The scientist knew that human hair is 5 percent sulfur by weight and contains no phosphorus. But the neutrons coming out of the rendered reactor top would have produced phosphorus 32, a radioactive substance with a known rate of decay, losing half its radioactivity in fourteen days. By determining the relative intensity of the phosphorus clinging to the hair samples, Petersen could approximate where the men were and make guesses about what they were doing based on the path of radiation up, down, and across their bodies.

Between Petersen's findings, Lushbaugh's three-dimensional study of injuries, and the technical team's analysis, a compelling picture of what was happening on top of the SL-1 reactor just before 9 P.M. on that January night was finally coming together. Dick Legg's crew appeared to be running behind in its work. The men were supposed to connect four of the reactor's five control rods; the fifth was a dummy that had remained bolted in place over the Christmas shutdown. But three hours before shift change, the crew was still working to latch its first control rod, the central one, designated number nine. The central rod had already been eased four inches out of the reactor core and had a C-clamp attached to hold it in place. One of the crewmen had then spun the lifting tool off the top of the rod so that a colleague could install a washer and nut, connecting the rod to the drive motor. A crewman had then put the lifting tool back on. At 9 P.M., chief operator Legg squatted down on the reactor top, straddling the rod motor for control rod seven. The position put him close to the bottom of control rod nine, where he could easily slip off the C-clamp. Jack Byrnes stood next to the port that housed the central control rod, in a position to gently ease the rod up to relieve pressure on the clamp. Trainee Richard McKinley was just off the reactor top. He was turned toward the other men, with his left leg and arm nearest the reactor.

At 9:01 P.M., Jack Byrnes pulled the hundred-pound central rod upward, strong and hard and in less than a second. The poison cadmium on the bottom end of the control rod—the barrier that had put the reactor to sleep two weeks earlier—slid out of the uranium fuel cells deep in the reactor's core. Like a hive of killer bees, neutrons in the uranium atoms swarmed instantly, colliding with one another, creating even more neutrons. In nanoseconds, there was an uncontrollable chain reaction which created enormous heat. At 3,740 degrees Fahrenheit, the uranium 235 fuel vaporized and

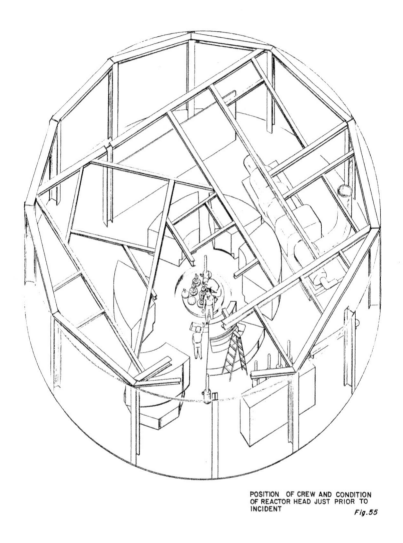

POSITION OF CREW AND CONDITION
OF REACTOR HEAD JUST PRIOR TO
INCIDENT
Fig.55

Diagram taken from the official interim report.

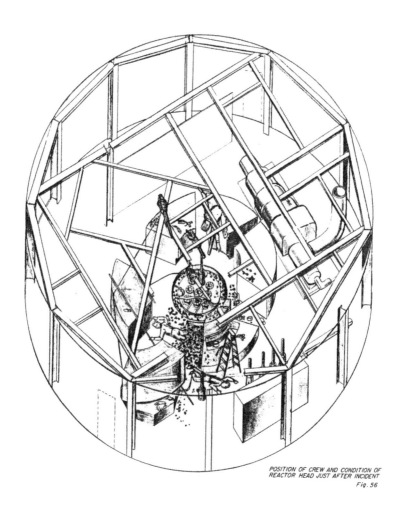

POSITION OF CREW AND CONDITION OF
REACTOR HEAD JUST AFTER INCIDENT

Fig. 56

created a fist of steam deep in the core. The steam rushed upward at tremendous velocity, aided by a two-foot void in the core—the water level had been dropped two feet over the holidays to aid in maintenance procedures. The water hammer slammed into the top of the heavy steel reactor top like a punch from a champion heavyweight. The blow lifted the nine-ton reactor core nine feet into the air, shearing off water lines that carried cool water in and radioactive water out, and spewing gamma, beta, and alpha radiation. Control rods, shield plugs—anything covering the ports of the reactor—became deadly missiles and shot upward at near-supersonic speed, powered by the rush of water and steam.

By the time the reactor vessel slammed back down into its hole, Legg had been impaled in the stomach and chest by control rod seven and hurtled into the ceiling. The autopsy report says death came "instantaneously from the destruction of his viscera by rapidly expanding gases that penetrated his abdominal cavity along with a heavy missile." Byrnes had been thrown back into one of the concrete shielding blocks, breaking ribs that pierced his heart. McKinley had been struck hard in the head by a piece of flying shrapnel, had his face torn off, and his left hand ripped from his arm. He lived for two hours, but he was in deep shock and never regained consciousness.

"It was all over in a matter of microseconds, certainly microseconds," says physicist Condit. "Those guys would never have known. That shock just happens…like that. It's all over before you see it."

And at 9:02 P.M., it was all over indeed.

The reenactment of that hellish second seemed to answer who was where, what they were doing, and what initiated the explosion. It seemed to further absolve Legg, the prank-ster boss, from a clearly not funny nuclear faux pas. But it implicated Jack Byrnes, the angry, distraught, disillusioned operator. The scenario made sense, technically and spatially. But it failed to explain one-tenth of a second that January night: What went through Jack Byrnes's mind in the moment before he tensed and pulled? What was his motive?

Bills says that when his technical group and the autopsy team presented their findings to the investigation commit-tee, there was an immediate response from its members, a response that would augment the wild rumors already cir-culating within a small group of nuclear workers: "Although they [the committee members] had done some investigating beforehand into the relationships of the people, the AEC's Division of Inspection went back again and really penetrated to see if they had missed something in those relationships that would explain violence, or maybe psychological things that were going on."

By that time, the committee had received off-the-record information about the crewmen's personal lives, as well as having heard some stories—unsettling ones. Those stories prompted them to investigate a potentially sensational angle to the story, one that could be summed up in three words: crime of passion.

*　*　*

Unbelievably, the world has never discovered—at least not from the committee—what its members learned about the crewmen's personal lives or how they thought that informa-tion may have impacted on what happened inside the SL-1 silo. Twenty-one months after the nuclear deaths, the investigation committee released its final report. It condemned the supervisors of the SL-1 project—from the civilian contractor to the military to the Atomic Energy Commission itself. But it said that ultimately these supervisory bodies were not to blame for the explosion. The committee pointed out serious operational problems that had been allowed to fester, including the loss of boron and the sticky control rods. But it said that those conditions had not precipitated the nuclear excursion. It found the training of the SL-1 crews seemingly inadequate. But it said lack of training did not play a role in the blast. It deemed the design of the reactor—its ability to go critical with the pull of just one rod—flawed. But it said that didn't lead to the deaths of the three young men.

The committee's report never mentioned the crewmen by name, nor did it identify who was doing what when the reactor exploded, nor what injuries they sustained. The report did not address the servicemen's personalities or psychological states of mind, their work capabilities, or their home lives. As persons, they were missing entirely from the document.

Instead, the board penned just two vague paragraphs that would have to pass for an acceptable official answer to the SL-1 mystery:

> The direct cause of the incident clearly appears to have been the manual withdrawal by one or more of the maintenance crew of the central control rod blade from the SL-1 core considerably beyond the limit specified in the maintenance procedure.
>
> The reason or motive for the abnormal withdrawal is considered highly speculative, and it does not appear at

all likely that there will ever be any reason to change this judgment.

In a cover letter on its final report, committee chairman Curtis Nelson offered just a few tantalizing words about what might have been at the root of the explosion. Rarely have so few words hinted at so much but actually said so little. The sudden pull on the control rod, he wrote, may have been motivated by one of two things:

> . . . involuntary performance of the individual manipulating the rod as a result of unusual or unexpected stimulus, or malperformance motivated by emotional stress or instability.

7

Murder-Suicide?

The Atomic Energy Commission must have thought the final investigative report a paper version of the concrete tombs that encased the hot bodies of the SL-1 crew. The report contained page after page of supplements, technical analysis, records of control rod drops, and nuclear jargon; the sheer weight of words formed an impregnable shell around the atomic incident. With the finding that the central control rod was pulled up too far and too fast by a crewman, the investigators closed the case. It was end of story: no names, no motives, no explanations. The media dutifully ran short stories on the committee's cryptic final statement and then moved on to other news. Lawmakers stopped asking questions. The twisted reactor core was trucked from the Hot Shop and buried with the rest of the hazardous waste. Grass grew over the layers of lead and concrete that surrounded the remains of Jack Byrnes, Dick Legg, and Richard McKinley. Their wives had scattered across the country. They were left alone with grief, toddlers, paltry military-widow benefits, and, soon, the knowledge that people were gossiping about them. It was that cryptic phrase in the final report—"involuntary performance... as a result of unusual or unexpected stimulus, or malperformance motivated by emotional stress or instability." To the families and friends of the victims, it was a whispered smear. To others, even cloaked in bureaucratic language, that one line in the final report hinted at dark things. Incredibly, no one at the time asked what the phrase really meant. It would be nine years before a government official would off-handedly provide the closest thing to an answer, and it would be eight more years before the public heard about it.

* * *

In September 1971, Stephan Hanauer settled in behind his desk at AEC headquarters in Washington, DC, to write a short memo to his boss. The professor-turned-bureaucrat knew his superior on the nuclear regulatory staff was looking into the threat sabotage activity posed to the nation's commercial nuclear reactors. Hanauer wanted to remind his boss to look closer afield than European terrorists or Russian spies. As Hanauer remembers it, he was thinking: "It isn't only shaped charges or guys with machine guns you have to worry about wrecking a nuclear plant. It's the people who work on it every day that you ought to be worried about."

The short message pointing out that danger was forwarded to his boss, who read it and filed it away along with the other official memos he received from the myriad bureaucrats who worked under him. The memo was no big deal, Hanauer says; it didn't raise a single eyebrow when he wrote it.

"I was trying to get people to talk about what really might happen instead of some of this spook stuff," Hanauer says. "I didn't get any heat for that memo. It was just part of the background."

Eight years later, in March 1979, the memo, yellowed with age, had been forgotten, just like the SL-1 incident. However, one man who had read the memo years earlier, nuclear regulator Robert Pollard, never forgot its contents. Pollard eventually had a change of mind about the worth of nuclear energy and left the AEC to join the Union of Concerned Scientists. During the height of anti-nuke opposition in the late 1970s, to buttress the group's arguments that nuclear reactors were unsafe, Pollard slipped a copy of Hanauer's memo into the hands of the editorial team at the *Brattleboro Reformer*, a small Vermont newspaper.

The paper ran the contents of the memo, which were quickly picked up by the major news services and distributed to hundreds of newspapers nationwide. Hanauer's memo was short in both length and detail. But it seemed to answer the question investigators had left hanging almost two decades earlier. And what an answer it was.

The explosion at SL-1 in 1961, Hanauer claimed, was no accident. It was sabotage—and it was committed by one of the three crewmen on duty that January night. Not only that, Hanauer wrote his boss, the first fatal nuclear reactor accident in the world was, in fact, a murder-suicide.

Some newspapers, probably fearing libel lawsuits, chose not to report why Hanauer believed one of the young servicemen would kill himself and take his comrades with him. But other papers did. Hanauer, they reported, believed the three deaths by atom, surely the most novel in the tawdry history of murder-suicide, were the culmination of a love triangle between the crewmen and one of their wives.

The revelation was shocking, with its suggestion that the most powerful force in nature could be wielded like a Saturday night special. But it hardly made a blip on the nation's radar screen. The press and public didn't seem to care; the story was passé by the next day. The Atomic Energy Commission made no formal statement. The SL-1 explosion, after all, was old news. By 1979, Americans found themselves desensitized to the horrors of senseless death and destruction—and government whitewashes. In the nearly two decades since the atomic disaster, they had lived through the dramatic cultural, social, and political upheavals of the 1960s and '70s. They had watched on television the assassinations of John F. and Bobby Kennedy and Reverend Martin Luther King Jr.; the aftermath of the My Lai massacre in Vietnam; the chaos and repercussions of the race riots; the shooting of students at Kent University; the heinous rampages of serial killers; and the disgrace of Richard M. Nixon.

Hanauer's memo didn't go completely unnoticed, however. The families of the dead men had sought anonymity after the explosion, and they didn't call any press conferences to counter the allegations. But privately, they were disgusted. They told friends that there was no truth nor proof to Hanauer's allegations. They said it was all a lie, just another insult, like the invasive autopsies, the strange burial rites, and the spineless final report by SL-1 investigators. Former cadre members in the Army Nuke program—whether they had known the three SL-1 victims or not—cursed Hanauer for sullying their colleagues, young men whom they believed had given their lives in the line of duty. Even those who had been part of the initial rescue mission were irked by the suggestion of scandal, perhaps feeling that it somehow made the extreme risk they had faced seem unnecessary, even foolish in retrospect. One worker publicly expressed how disappointed he was with Hanauer, feeling the AEC official had exercised poor judgment in writing the memo. He said that Hanauer, who hadn't been anywhere near the SL-1 site and hadn't known the men personally, should have used more discretion.

If the memo went over the heads of Americans, its contents would become an urban legend in the closed world of the nuclear industry. The love-triangle story would pass from old hand to new, embellished here, spiced up there. Facts had always been scarce, and names could never be attached definitively to the alleged sins and crimes. Nonetheless, retelling the strange story of SL-1 became salacious amusement, a lunch-hour whodunit, even a cautionary tale to new nuclear workers.

After the memo was published and the love-triangle story was circulated, embellished, and then hardened into legend in the nuclear world, Hanauer's name would be cited as the informed source. Now in his mid-seventies and working with the Department of Energy, Hanauer wishes it weren't so. It's not that he doesn't think sabotage couldn't happen at an American nuclear reactor, or that his instincts aren't telling him something strange happened at SL-1 that night in 1961. It's simply because today, as on that day in 1971 when he wrote the memo, he has no proof. He didn't talk to investigators. He didn't talk to family members. He didn't hear a remorseful wife confess to an affair. He found no angry suicide note. In fact, Hanauer now admits his memo was really not much more than the written version of someone leaning toward the person beside them and saying, "Hey, did you hear about...?" The memo was important, though, because it was the first time a nuclear official reported, in writing, the kind of talk that had been circulating among a small group of insiders since 1961. And if the memo was based on rumor, at least it came from sources close to the men involved.

"I know exactly when I first heard it. I lecture at MIT [Massachusetts Institute of Technology] every summer in their nuclear power safety course," Hanauer says. The first time I did it was in 1966, and some people from the army reactors program were in the course. They told me about this love triangle in such a way that I believed it. They convinced me at the time that they knew what they were talking about. They were convincing.

"But, of course, I didn't cross-examine them. I listened. I don't remember any details, and I was less critical then than I am now. I don't know who they were, and I don't have any evidence."

And as it turns out, Hanauer's take on the events of January 3, 1961, may not have been entirely ungrounded. An investigator would later learn that passions—maybe petty, maybe deep—were in the silo that night. But could anger, sorrow, spite, or jealousy really have prompted Byrnes to abandon his training, his common sense, his conscience? Who knows? Whatever emotion was running wild at 9 p.m. was extinguished by 9:01. It left investigators to divine by clues, elliptical comments, and recollection what role passion might have played in America's first reactor deaths.

*   *   *

Arlene Byrnes opened her door in the early hours of January 4 to find a group of grim-faced people congregated on her doorstep. Given the odd hour and the somber faces in front of her, Arlene knew they were not bearing good news. Her first thought was that Jack had been hurt in an automobile crash. That was understandable. Highway 20 heaved and fell with the desert's contour. In the winter it was pummeled by ferocious winds and driving snow, which collected in its swales. It could be treacherous, and Jack wouldn't have been the first site worker to be hurt or even killed on the ribbon of asphalt that led to and from the Lost River Desert.

Once she had been escorted into her living room and guided to a chair, however, army officials told Arlene her husband had been killed in an accident at SL-1. Her response, between the tears and shock, was odd. She turned to the army captain, sergeant, and wife of another sergeant huddled on the couch and said Jack had told her—she didn't say when—that the reactor would blow up and he would be killed in the explosion. He even instructed her to have one of his good friends, a civilian employee at the reactor, invest her money if something happened to him. It was a peculiar statement. But those present didn't question the distraught widow. Maybe they thought that in her traumatized state, she was misinterpreting or exaggerating some comment Jack had made. Maybe she was implying—and this seemed most likely—that Jack believed the reactor was unsafe. But there was another way of interpreting her statement, and the implications were staggering. Could Jack's warning have been a veiled threat—or a distraught promise?

In the coming days, the investigation committee quickly appointed by the AEC would ponder that disturbing question, but secretly. Arlene's late-night revelation and another bit of information—that a crewman's wife had made an unsettling phone call to a top site official—would quickly lead investigators into murky psychological waters rarely navigated by the orderly minds that dominated both the AEC and the nuclear industry. The scientists and professors on the committee knew the half-life of every radioactive element known to man. They comprehended complex mathematical formulas. But the human heart? Too messy, too unpredictable. It soon became clear committee members were learning about human fallibility, and it was not a lesson they wanted to share with the public.

On January 16, Allan Johnson, manager of the AEC's Idaho Operations Office, issued a memo that for years afterward would guide what the public learned, and what it wouldn't: "All documents concerning the SL-1 incident, not otherwise classified Confidential or higher, are to be marked Official Use Only. Exceptions occur in the case of information published in the form of press releases. Press releases themselves are Official Use Only until approved by the Manager or his designated alternate(s)."

Even site medical director Dr. George Voelz wasn't privy to the details of the investigation. But as one of the higher-level employees, he was aware that something was brewing: "I answered questions from the investigating committee, two or three or four days afterwards. They were meeting all day long with various people to try to get details on what happened and what the status was out at the SL-1 site—the psychological things I wasn't directly involved with. But I knew that there were questions being asked. This got started within about twenty-four hours. There was a telephone call between one of the crewmen and his wife. They'd been having problems over the weekend. He was out at the site, and I guess the telephone call didn't go very well, and he got angry and slammed down the phone. And a very short time later, this accident happened. Then, soon after, one of the wives had actually called the [site] manager and expressed her concern that her husband may have done something rash. And that triggered off immediately this whole investigation about what was known about the stability of this crew."

Then site manager Johnson is now dead, and records don't indicate which wife called him. But within days of the accident, Leo Miazga, a special investigator with the AEC's Division of Inspection, then working in Richland, Washington, was flown into Idaho Falls to conduct classified interviews—and he was told to focus his inquiry on Jack Byrnes. Even though months would pass before the investigation committee received confirmation from the autopsy team that Byrnes was the one pulling the rod that night, its members saw evidence mounting that twenty-one-year-old Jack Byrnes played a key role in the incident. They'd established that Byrnes was having problems at home. They knew he had received a troubling call from Arlene shortly before he died. They had a record of a second call that didn't get through, followed by a frantic call to the operator that something must be wrong at the reactor. They had learned of Jack's comment to his wife about the reactor exploding. They must also have heard rumors about Byrnes's sex life.

It all smelled funny, but did it mean anything? Armed with only that scant information, AEC investigator Miazga got down to work, talking to anyone who might possibly have information about Jack Byrnes's personal affairs. Surprisingly, a gas jockey was one of the first in the nation to know there might have been something kinky going on at SL-1.

Homer Clary was instructing beginner skiers at Taylor Mountain, southeast of Idaho Falls, a few days after the accident when a couple of men in suits showed up, looking out of place at the bottom of the ski hill. The investigators, one of whom was Miazga, had heard Clary had been at a party during the holidays with some members of the SL-1 crew and had skied with Byrnes just a day or two before the explosion. They also knew Clary worked with Byrnes occasionally at Kelly's Texaco Station. Based on the kind of questions the men asked and a few of their comments, Clary reckoned something was up.

"The CIA-type people who questioned me shortly after the accident were asking me questions about whether there was something going on within the group of the three guys, problems with their wives," Clary recalls more than forty years later. "They were speculating that one of the guys was fooling around with the wife of another guy. I don't remember who they thought was fooling around. But I know they were quite heavy into this kind of thing. I had no information for 'em. I just know there seemed to be a lot of speculation on that area."

The only thing that Miazga's interviews with Byrnes's neighbors and coworkers was able to confirm with certainty was that Jack Byrnes was leading a troubled personal life, increasingly unhappy in his marriage and in the responsibilities he had taken on at such a young age. Neighbor Robert Meyer confirmed that Jack and Arlene argued frequently about money and her housekeeping, and that Jack spent very little time at the couple's duplex. Byrnes flitted from nightclubs to the Texaco station to the ski hills—he seemingly wanted to be anywhere but home. Meyer also said that he saw Byrnes show signs of temper at work when he talked to his wife on the phone, apparently not caring who overheard. Another Texaco station worker, Jim Meak, told the investigator that Byrnes "invariably got mad and used profanity in talking to his wife." Meak also recounted how one night Byrnes had asked that he cover for him should Arlene call while he was out carousing with a friend. Another neighbor of the Byrneses, Robert Matlock, talked about the couples' screaming matches—often punctuated by Jack's clothing being tossed onto the lawn—and how they were as routine as the mail delivery. He also confirmed that the couple had had a loud fight on New Year's Eve, a fight that prompted Byrnes to leave home and shack up with a buddy. Byrnes's friend Martin Buckley recalled how upset the reactor operator was about his marital problems and about his missing paycheck, which he thought Arlene had taken. He also admitted that he and Jack had downed a couple of beers an hour or so before Byrnes headed off to work on January 3.

By January 20, seventeen days after the accident, Miazga had prepared a classified memo for his boss back in Washington, DC, who forwarded it to Curtis Nelson, the investigation committee's chairman. The report was never released to the public, given to the victims' families, or circulated outside a small group of top government officials. Miazga, through seven interviews that covered seven pages, recreated the final days of Jack Byrnes. All in all, it was a succinct report of a rough three days for Jack and Arlene. It was a quick sketch of a young man who arrived at SL-1 on the night of January 3 upset, angry, and frantic about his domestic problems. However, Miazga apparently did not talk to Arlene directly; her insight into Jack's state of mind was missing from his investigation. And although Miazga had been asking around about Jack Byrnes's sex life, the report did not suggest anything unseemly.

What conclusions did the investigative committee draw from the short report? The members left no written record of their reactions to or thoughts on Miazga's investigation. But it's possible that they were simply fishing, casting nets to explore every possibility. At the time, they were only two weeks into their investigation; the technical team had yet to figure out how the reactor exploded, and the committee had just heard a lot of testimony that painted a picture of a reactor beset by mechanical problems and hands-off management. There were too many unknowns and too many suspects.

When the members of the investigation committee returned to Washington, DC, at the end of January 1961, a month-long silence followed. C. Wayne Bills, the head of the technical team probing the SL-1 explosion, says committee members were tight-lipped about what they were thinking, even after they learned that the explosion had occurred when Jack Byrnes was grasping the control rod.

"As far as I know, the investigating committee really shut up," he says. "There wasn't much that leaked out under the tent. And there was Nelson and his sidekick, the guy who did most of the investigations for him. I think they found there was some kind of hanky-panky going on, but I didn't know even which way it was. I think that was probably substantiated a little bit by some of the people in the cadre. I don't know that they [the commission] had any details, but I think it was more than rumor."

It was the persistence of rumors about "hanky-panky" that prompted Miazga to conduct a second round of interviews. In the summer of 1962—more than a year after the deaths—the AEC investigator was again dispatched to Idaho Falls to probe a little more deeply into the human dynamics at play between the men who had been working on the reactor when it exploded.

The report Miazga produced on July 25, 1962, implies ever so subtly that he was asking around about the rumor that Judy Legg and Jack Byrnes might have had some sort of relationship. In one paragraph, Miazga wrote about his interview with Sergeant Gordon Stolla, a chief operator at SL-1. Stolla recalled that Judy had been a stenographer at the Testing Station when she met Dick Legg. Obviously responding to a question posed by Miazga, Stolla's statement is paraphrased in the report as follows: "He said Miss Cole was respected by her fellow employees and he does not believe that any incidents took place which might cause friction among [SL-1] personnel."

Miazga's use of Judy's maiden name may or may not be significant; it may imply that Judy was unmarried when the rumored liaison occurred. That sentence is the only reference in Miazga's two reports and in thousands of pages of documents about SL-1 that mentions Judy or possibly hints at a love triangle among the men and one of their wives. But the absence of official confirmation of the love triangle doesn't mean that others weren't convinced of its existence. When interviewed for a 1981 documentary about the SL-1 incident, Charles Luke, a senior scientist at the Testing Station, was asked about the possibility that the explosion was the result of a sexual tryst. Staring into the lens of the camera with a wry expression, Luke delivered a brief, inconclusive take on the love-triangle theory, but it was pointed all the same. "Direct evidence? No. Hearsay? Yes," he said slowly and deliberately.

Another part of Miazga's second report was inspired by a tip about the bachelor party in May 1960 where, intoxicated with sex and alcohol, Jack Byrnes and Dick Legg ended the night swinging at each other. Somehow, word of the tawdry boys' night out had filtered to an army captain, and he had passed on the information to the AEC's investigation unit, which in turn notified the investigation panel. Miazga, from the kind of questions he asked, was undoubtedly instructed to probe the possibility that Byrnes and Legg held lingering grudges from their confrontation at the party.

Miazga's second report, like the first, was not released publicly or given to the victims' families. Obtained all these years later through the Freedom of Information Act, it bears the commonplace title "SL-1 Incident (Supplemental Report)." But there's nothing commonplace about it. It's the only evidence of what investigators meant when they issued their final statement two months later, suggesting that the nuclear catastrophe was the result of "malperformance." Miazga's twenty-four-page report is written in the same style as the first: plodding and with no direct quotes from interviewees. His questioning seems to meander, and the report gives no sense of how hard he may have pushed the people he interviewed. But the report holds a reader's interest because of what it reveals—and what it doesn't; what it suggests—and what it doesn't; and what the commission ultimately knew—and what it didn't. If there were a treasure map for the mysterious saga of SL-1, Miazga's report would mark the spot. It may be the wellspring of all the rumors and speculation that have made the SL-1 explosion such a perplexing mystery.

*   *   *

It is quite possible that the details in Miazga's report of the liquor-sodden bachelor party were the genesis of the love-triangle rumor. Participants recounted the night for the investigator: Jack Byrnes inviting Mitzi, the prostitute, to an after-hours party; Byrnes's indiscretion with the "woman of easy virtue," as Miazga put it; and the drunken confrontation between Byrnes and Legg soon after. What the participants described wasn't love. But the whole sorry affair had three actors in it—the elements of a triangle. Perhaps over the years, as the details of the report leaked and then traveled from person to person, the sex and violence at the bachelor party morphed into something grander.

It is much harder to explain the presence—if barely perceptible—of Judy Legg in that same report. Something clearly prompted Miazga to question those who knew Jack about Judy. Years later, Miazga allegedly told at least one person, a Salt Lake City–based documentary maker, that there was some substance to the love-triangle rumor. But he didn't say what it was, and he may have been posturing. He may have conveyed sensitive information to his superiors verbally, leaving it out of the report. Or he may have known nothing at all about an extramarital relationship and was just poking around when he asked Stolla about Judy. Government records repositories contain no other memos or reports from

Miazga that mention, even indirectly, a love affair in general or Judy Legg in particular. Leo Miazga died years ago and, according to a nephew who helped clean out the investigator's home after his death, no notes, letters, or documents were found that mentioned the reactor accident.

If Miazga's reports failed to shed light on the love-triangle rumors, they did provide fodder for those who suspected that a roiled heart, not a stable mind, was holding the SL-1's central control rod the night of the explosion. Remove any trace of indiscretion and there remains the possibility of suicide. And if the AEC investigators were looking for signs that psychological forces were at play, forces that would drive a man to such an extreme, Miazga's report gave them plenty. Interviews with more than a dozen reactor supervisors, co-workers, and neighbors painted a picture of two young men who seemed to be spinning out of control in the months leading up to the disaster.

The confirmed romp with Mitzi and other rumors about his love life may have said something about Byrnes's elastic take on his marriage vows. But of greater significance was the fight that ended the bachelor party: it gave a glimpse into the edgy interior worlds of Byrnes and Legg. Miazga's interviews with coworkers painted a picture of two impetuous hotheads who brought their anger to work. A string of supervisors described Byrnes as a lit firecracker, a guy with an internal rage that manifested in a flushed face, a fierce twitch in one eye, and airborne tools. He would arrive at work morose about personal problems or resentful that he hadn't been promoted to chief operator. Byrnes, witnesses said, made it clear he hated taking orders, disliked having to kowtow to military authority. He was argumentative with bosses, truculent even. He was a problem on the job, wanting to do things his way, even if he didn't have the experience to know what the right way was. He had an agenda, a schedule for advancement in the nuclear industry, and he made it no secret that it wasn't going as planned. Byrnes dreamed of getting out of the service, becoming a nuclear operator and eventually a supervisor at a nuclear plant then in the early stages of planning in upstate New York. If he couldn't prove himself at a small test outfit like SL-1, that dream could quickly disintegrate.

Legg didn't fare much better in most of the supervisors' descriptions. He too had a temper that he found hard to keep under control; his startling challenge to fight a superior officer during a Christmas party was clear evidence of that. Some also cited a lax work ethic. Legg avoided performing required building checks, he had slept in his car while on shift, and he covered for a buddy who left the reactor to go to town. At least one manager believed that Legg carried around a "little man's" chip on his shoulder. Like Byrnes, Legg was pugnacious and chafed at military authority. He didn't think twice about using reactor equipment as props for his pranks and practical jokes, and he found the SL-1 reactor silo a ready ring for his impromptu wrestling matches with coworkers. To make matters worse, Legg arrived at work on January 3 with his future uncertain. He was scheduled to meet with his boss the next day—the one he had challenged to a fight—to learn if he was to be demoted or even transferred out of the reactor program altogether. It was a stressful situation for a man whose child was soon due.

Those interviewed were divided on whether the two men arrived on their dying day holding a grudge over the alcohol-fueled fight six months earlier. Some said they didn't think so. The two men's moods and spells appeared as shallow and fleeting as the creeks that ran in the desert during the spring. Others thought both men quite capable of harboring hard feelings. Legg had a talent for goading people. Byrnes had a penchant for acting out. But neither had ever talked publicly about what had led to the fight, and people couldn't remember if they had worked together after that night in May 1960. Miazga could not find consensus among the interviewees about the impact the fight might have had on the working relationship between the men on the night of the explosion.

However, there was unanimous agreement among the supervisors that both men knew they should never pull a control rod out too quickly or too far. It was common knowledge, the supervisors said, drilled into each and every trainee. But would the young men have known the consequences of breaking the rules? The supervisors said they weren't sure. Even they expressed amazement at the cataclysmic damage inside the silo after the central rod was pulled out too far. Most thought it would have created nothing more than an intense radiation field as neutrons in the core went crazy. Others thought a sudden rod withdrawal would take the reactor critical, destroying the reactor core. Most assumed the sudden increase in heated water would be vented through the system's cooling pipes. But an explosion like the one at SL-1? It seemed unlikely that the young, relatively inexperienced Byrnes and Legg would have foreseen the consequences of a pulled rod, especially since the severity of the explosion was attributed to the void in the reactor core created when the water level was dropped for maintenance two weeks earlier. But one man interviewed, a civilian worker and Jack's best friend, recalled a discussion the two had had back in the summer of 1960. They were talking over coffee about what they would do if they were manning an SL-1–type reactor up in the Arctic and the Russians attacked over the polar cap. Byrnes told his buddy he would destroy the reactor by pulling the central control rod.

But would either man have actually done it, actually yanked up on that critical control rod? Miazga asked those he interviewed. One sergeant said Byrnes just might have if goaded by Legg. On that January night, Legg could have pushed Byrnes too hard—made some crack—and in retaliation, Jack may have pulled the central rod in order to foul up the shift's work and make Legg look bad. Another sergeant said Byrnes was an inquisitive kid and might have yanked up the rod just to see what would happen, never imagining the consequences. The wife of another sergeant at the reactor thought that young Jack Byrnes was an impulsive guy: he drove too fast, lived too hard, did things without thinking. She thought it was possible that he pulled the rod on a whim. But she also speculated that, given his personal problems, he could have done it deliberately to take his own life. She didn't think Legg would have done so. He was cocky and conceited, and she said people like that would never do something that would harm themselves or lead to self-destruction.

Taken in its totality, Miazga's second report seemed a damning indictment of Jack Byrnes and, to a lesser degree, Dick Legg. But nothing about the whole sorry saga of SL-1 would prove simple and clear-cut, and Miazga's report was no different. Some of those interviewed said they had seen other, less volatile sides of the two men. In those accounts, the two didn't seem to be anything but what they were: two guys in their early and mid-twenties with a lot of testosterone, ambitions that exceeded their experience, and a zest for fun, whether it was pranks or skiing or going out drinking with the boys.

Jack Byrnes's best friend in Idaho, Robert Young, said he'd been camping, hunting, and skiing with Jack and had never witnessed the kind of temper described by reactor supervisors. He said Jack was a great guy to hang out with. Although Brynes sometimes expressed frustration with Arlene, he appeared to love his wife and his little boy. Young said Byrnes did dance with unaccompanied women when he went to clubs without Judy. But he said Byrnes, to his knowledge, didn't have a liaison with any of them—for the most prosaic of reasons: "The level of army pay does not permit any high living or extramarital affairs," Miazga paraphrased Young as saying.

The same supervisors who damned Byrnes's unmilitary-like attitude conceded that he possessed above-average intelligence and seemed sincerely interested in mastering the operation of the reactor and getting ahead in the industry. He did good work on the things he was qualified to do; he just had to be left alone to do it his way. The same woman who said Byrnes was reckless also said he was charming and had always treated her politely. One sergeant mentioned Byrnes appearing to be under intense pressure on the last day of his life, but another said he spoke to Byrnes on shift change and he seemed fine. Miazga discovered Byrnes had taken out a life insurance policy on himself. But as it turned out, the purchase had been made six months before the accident, and the insurance agent had been forced to pay Byrnes a couple of visits before the soldier agreed to sign on. And the payout was a paltry $2,240, double if Jack died accidentally. The premium was $5 a month.

The same sort of conflicting character evidence clouded the investigation of Dick Legg. Like Byrnes, he sometimes had an attitude at work, superiors said. Furthermore, he obviously had exercised poor judgment in covering for his friend who wanted to take the day off, as well as in some of the pranks he pulled. But otherwise, they said, Legg seemed competent. One supervisor, in fact, said Legg had a reputation for pushing his crew to do the work right and on time. As for Legg's personal life, Homer Clary, the part-time worker at the Texaco station, said he had had dinner with Dick and Judy during the Christmas holidays and they seemed happy. He and Dick chatted about archery. Legg struck Clary as a good guy. Another supervisor said that on the day of the explosion, Legg didn't seem troubled and didn't mention his meeting the next day with the sergeant he'd challenged to fight. In fact, they chatted a bit about Judy and how she was thrilled to be having the couple's first child in the coming month. Miazga's report did not mention Dick and Judy's private affairs. He either found nothing worth reporting or he conveyed sensitive information verbally to superiors, leaving it out of the report.

What did the investigation committee make of Miazga's second round of questioning? Were Byrnes and Legg loose cannons? Or were they simply high-spirited enlisted men with typical problems? Committee members left no written records of their deliberations. But by the time they received Miazga's final report, they had the autopsy findings as well as the conclusions from the technical committee. They believed the major findings in both: the reactor exploded because a crewman had withdrawn the central control rod too far, and Jack Byrnes was on the central control rod when the reactor blew. If technical problems such as a sticking control rod or the loss of boron in the core were ruled out, only a narrow number of possible scenarios were left.

One: It was simply an accident. Maybe Jack Byrnes was daydreaming on the reactor top or preoccupied with his troubles, and he simply raised the rod without thinking. Byrnes had often seen control rods pulled out a couple feet when the reactor was up and running. Maybe he acted on that memory rather than his bosses' warnings about raising the central rod too high when the reactor was shut down. The only drawback to that explanation? Byrnes had to raise the rod only a quarter of an inch for Dick Legg to remove the C-clamp.

Two: Dick Legg really did playfully grab Jack Byrnes at the wrong time. Jack would have been in a vulnerable position as Legg crouched down to remove the C-clamp. And everyone knew Dick liked a good goose. It was still a remote possibility, even though the technical team hadn't been able to get volunteers to jump high enough and withdraw the central control too far with their mock goosing.

Three: Dick Legg said something that infuriated Jack Byrnes. Legg may have known that Byrnes had left his wife and might have overheard the emotional phone call between Jack and Arlene earlier that evening. Perhaps he said something malicious, connecting the breakup with Jack's indiscretion with Mitzi. He could even have made a crack about Jack's performance during that indiscretion. Legg could be quick with his tongue, but would he have been that cruel?

Or Legg might have criticized Jack about his work that night—they were running behind schedule, after all. Dick was certainly capable of goading his crew, and Jack didn't like to be goaded. In response, Jack could have decided to jerk the rod to get Legg's attention, perhaps to spoil the night's operation, or with the hope of getting Legg into trouble. Maybe he just yanked it as the ultimate retort to whatever Legg said to him, without a thought to the consequences. Like Legg, Byrnes had a history of being quick to anger.

Four: Jack Byrnes, already upset about his marital problems when he arrived at SL-1 that night, could have been fuming after the phone call from Arlene. She had told friends the couple had decided their marriage was over and had talked about what was going to happen to Jack's paycheck, the one he was agitated about not finding in the mailbox earlier in the day. He could have been roiled with emotion: anger, remorse, guilt, feelings of persecution—all the high, wild feelings that often hit when a marriage disintegrates. Maybe he wanted to do something to get Arlene's sympathy, or just to "show her." Or, maybe, in that split second before he pulled, an emotional Jack Byrnes decided the future was just too grim. The kid who was always in a hurry to grow up pulled—and pulled hard—to nullify it all.

It was more of a tangled, mixed-up affair than a group of middle-aged, eastern bureaucrats were willing to deal with. It was 1962 and psychological profiling was primitive. All the possible scenarios seemed outlandish, each in its own way. The variables at play in the SL-1 explosion were the dramatic building blocks of a Hollywood film script, not of a final report about a military nuclear disaster. The nuclear industry hinged on predictable science and logic and was populated by sensible people. Choosing any of the human scenarios as an explanation for the SL-1 accident—especially when they were all based on speculation—must have been distasteful to the committee members. All the scenarios implied that the nuclear industry had put the public's safety in the hands of an angry, brokenhearted kid with little real experience. Certainly the committee members must have known that it wouldn't do to settle on any one of those four scenarios as the definitive cause of the world's first nuclear reactor deaths; all of the choices were lousy.

Six weeks after receiving Miazga's chronicle on the life of Jack Byrnes, the committee released its final report to the AEC and the US Congress. There was no mention of drunkenness, illicit sex, radiation-sterilized flesh, headless torsos, or concrete graves. There was a *de rigueur* reprimand to all the managers of SL-1 for shoddy management and use of a questionable reactor design, but this was qualified by a declaration that those factors didn't directly cause the horrible incident. There was a suggestion that the crew's training was perhaps inadequate, but not sufficiently so as to be responsible for the accident. Readers looking for a cause of the explosion would have to settle for the cryptic phrase: "involuntary performance...as a result of unusual or unexpected stimulus, or malperformance motivated by emotional stress or instability." It was, perhaps, as some have charged, a finding that said nothing. But it hinted at everything, conceding strange forces may have been unleashed in the Lost River Desert. That concession—that nuclear energy was only as safe and sane as the humans wielding it—had to have been galling to the investigation committee.

Years later, before his death, committee chairman Curtis

Nelson made just one publicly recorded statement on what investigators really thought about the whole ordeal. "We were unable to fix a real definite cause," he said. "It couldn't have happened, yet it did. It shouldn't have. We talked to anybody and everybody who, in our minds, could possibly contribute anything, including relatives and friends.

"One of the boys had had some very bad news of some family sort. And it was our rather far-fetched guess that maybe he wanted to kill himself—or didn't care what happened, more likely."

*   *   *

With the passage of four decades, there are few people left alive who were in the desert the night SL-1 exploded, and fewer still who want to talk about it. But of those who are alive and will talk, the mystery of what caused the nuclear blast remains intriguing, a rumination on the complexities of human behavior.

Richard Lewis, then an air force master sergeant and the superintendent of SL-1, remains baffled after all these years. He can't wrap his mind around the idea that one of his crewmen blew up the reactor intentionally: "It could have been done deliberately for whatever reason, but it doesn't make any sense to me," he says.

The man who headed the search for a technical cause for the explosion, C. Wayne Bills, thinks a moment of pique might have precipitated the incident, but he is resigned to never knowing for certain: "I think that for some reason, you just get the guy—whatever his psyche was at the time—who just said 'To hell with it' and jerked. I'm not sure it was as premeditated as some think. They may have been talking to each other. You don't know what kind of dialogue was going on right at that time. Why that happened and the stability of the guys... I just figure it's all in that mystery zone somewhere."

Historian Susan Stacy was one of the last people to talk to John Horan, a longtime influential site manager, before his death. Horan was the AEC's director of health and safety at the time of the SL-1 explosion, and Stacy says he was privy to a lot of confidential information, as well as to the workings of the investigative committee: "When I talked to John, he believed that emotional instability, as he put it, contributed to this accident. He felt that one of the men wanted to obtain the sympathy of his wife by injuring himself, which is a sick thing to do but you know you've seen that scenario in many families' histories. And he thought maybe the guy tried to do something that would injure himself but would not kill everybody, something to get his wife back into the business of worrying about him, to having sympathy for him... I don't know, I think that's a sophisticated way of thinking about getting attention. But why John Horan, who was a man of tremendous experience and was in a position to know, believed that, is the thing that sticks in my craw."

Dr. George Voelz, the site's medical director at the time of the explosion and one of the few men in the world to see the effects of nuclear death up close, says he and Lushbaugh, the doctor who did the autopsies, often discussed the essential mystery of SL-1, and they disagreed on the catalyst. "I have concluded—and this is different than other people feel—it was a suicide and he took a couple people with him," he says. "I talked to Lushbaugh about it in earlier years, and he did not feel it was [suicide]. But no one else has given another mechanism that had any significant probability of happening any other way. When you just put two and two together with all the other things that were going on in this group and the hearsay that I had gotten in regard to some wife calling in and suggesting her husband did something rash, I've just come to the conclusion that was the highest probability.

"You know, ultimately the cause for this thing—if you go back a little deeper—was that about eighty percent of the control of this reactor was on that central rod. In talking to some nuclear designers, I asked, 'Why put so much control on the one rod?' 'Well,' one of them said, 'it just made things simpler.' 'But,' I said, 'you've left the possibility that someone could really pull that central rod and you could get a burst of energy.' He kind of looked at me and said, 'Yeah, but no one in their right mind would do that.' And I said, 'Well…that's one of the possibilities. We see people every day who aren't in their right minds.'"

8

Nuclear Legacy

More than two decades after the SL-1 reactor explosion, officials at the Testing Station decided to tear down the lone structure still standing at the site. The white, two-story administrative building had been decontaminated after the incident. Everything inside was painstakingly cleaned by hand and put back into circulation. Fluorescent lights were taken apart, decontaminated, and reassembled; furniture was dipped into detergents and wiped clean; wrappers on office supplies were discarded and their contents saved; even the candy vending machine was washed and restocked. The building, with new interior walls and floors, was used for several years afterward, then abandoned when the army shut down its nuclear program and the site managers couldn't find any other uses for it. It squatted in the middle of the Lost River Desert for a decade, nothing more than a weathered, unmarked memorial to a macabre and mysterious event.

Doug Caldwell, a radiological control technician, helped dismantle the office building in 1993, chip up asphalt, and box low-level radioactive debris found in the yard. In the dead of winter, when the sun set early and the desert was plunged into blackness, Caldwell came to believe—or at least he said he did—that the ghosts of the three men killed in the explosion roamed that patch of sagebrush. "When it was night, you could swear you could see them looking at you through [the windows of] the old buildings," Caldwell told a reporter for a news story published in the *Idaho Falls Post-Register* in 1995.

The image conjured by Caldwell seems a nice, if trite, end to the saga of the SL-1 explosion, with its suggestion that the unsettled spirits of the dead crewmen roam the Lost River Desert looking for vindication and an end to the rumors that have tarnished their honor. But that's all fairy dust. The reality is Byrnes, Legg, and McKinley are encased in lead, locked in caskets, stored in vaults, and trapped by concrete. They're dead, and they'll be adding nothing to the technical evidence, partial truths, and rumors.

But there are people willing to speak for the two long-dead men and their hapless cohort, people who steadfastly believe the crewmen deserve exoneration. They have an alternate vision of that night inside the reactor, one that doesn't have Jack Byrnes, in a moment of anger or despair, tensing his muscles and pulling hard. Their version of the SL-1 story is about duty, sacrifice, and innocence.

The young men's families and most of their colleagues have always believed that none of the crewman did anything wrong on the night of January 3, 1961. They say the explosion was simply an accident: neglected technical problems collided with a poor reactor design and things went boom.

Some believe Jack Byrnes might have died from being too conscientious. They suggest that after Byrnes and Legg con-

SL-1 Reactor Transfer Operation

The SL-1 reactor is transferred out of the testing grounds.

nected the central control rod, Byrnes might have decided to "exercise" it—move it up and down slightly—as all crews had been instructed to do at the start of each shift as a quick fix for the problem of sticking rods. Some of the Army Nukes believe that because of the frigid temperatures that January, icy water was circulating in the shutdown reactor and, because of the previous maintenance work, at a lower level than usual. That change in the reactor's core temperature, along with the loss of the poisonous boron, was enough to put the nuclear machine on the verge of going critical. And that, they say, is exactly what happened when Byrnes raised the rod slightly.

If their version of events is true, if the explosion was simply an industrial accident, then the ghostly faces that Caldwell says he saw peering through the building's windows on dark nights would be those of perplexed young innocents, offered up as a sacrifice by an industry more interested in covering its collective butt than in taking responsibility for a mismanaged reactor program. And the clandestine rumblings about a love triangle leading to murder-suicide? The investigation commission's broad hint that stress or instability led Jack Byrnes to do something terribly wrong? The suggestion that the crew was undertrained and perhaps underscreened? All just a convenient and effective smoke-and-mirrors campaign to deflect attention away from the reactor's obvious flaws and its managers' sloppy performances. At least that's how some people view the official response to the explosion.

"The Atomic Energy Commission really worked hard on putting a smooth coat on the whole thing," says Clay Condit, the physicist who monitored the investigation for the navy. "The AEC was providing the money and it was supposed to be responsible [for overseeing the SL-1 project]. They were aware of the technical problems, but they didn't look into them closely. [Politically, the investigating committee] could not have said the AEC should have shut that facility down.

"It was really nice to say, 'Hey, these guys, it was their fault. They had this love affair going or one of them committed suicide.' That's how this bullshit started. It was, I think, one hundred percent fabrication. Somebody threw that out on the table and they said, 'Let's go with that sucker.'"

Stephan Hanauer, the nuclear safety expert who first publicly raised the specter of a passion-provoked murder-suicide, remembers top nuclear officials scurried like rats following the explosion: "The Joint Committee on Atomic Energy in the US Congress was very, very powerful. And after the accident, they held a hearing [in June 1961]. The chairman said, 'Who is responsible for this dreadful accident?' And a whole bunch of powerful people said, 'Not me.' Finally, some captain or major in the army stood up and said, 'I was in charge, and it's my responsibility.' And the chairman would have none if it. He said 'Yes, yes, of course. But, in fact, the responsibility is much broader and much higher.'"

Hanauer was working at the government's Oak Ridge Laboratory when the explosion occurred. Long before he heard the rumor of a murder-suicide, he thought the explosion needed to be examined by the nuclear industry for lessons. His superiors apparently thought otherwise: "I actually wrote an article about it. I was going to publish it in *Nuclear Safety* magazine, which was an AEC publication, and it was rejected because it was too critical of the [reactor] design. It was rejected in Washington."

More than four decades after the catastrophic event at SL-1, Hanauer still suspects that something peculiar happened between the men on shift that January night in 1961. But even if that's true, there's little doubt in his mind that the reactor's managers put a loaded gun into a young guy's hand.

"We'll never know if one guy pulled a rod on purpose or if he knew it would blow the plant up or if that was his intention or if it was something else," he says. "We'll never know any of that. But the plant ought to be resistant to the more obvious schemes."

Hanauer admits he, too, learned some lessons from the SL-1 affair. He now regrets his own role in promulgating the rumors that tainted the reputations of SL-1's military crew and brought pain and further suffering to their families. "People periodically bring the memo up, either to give me trouble or somebody else trouble," he says. "I'm the one who put it [the love-triangle theory] on the map, and I'm sorry. I don't think I would say that about three guys today unless I knew it."

Ed Fedol was one of the hopeful young servicemen who came to SL-1 in the early 1960s looking for a promotion and chasing the bright future the atom seemed to promise. He had to rush to the East Coast two weeks before the explosion for a family emergency, but he still carries an image of Dick Legg: "He was kind of a fun-loving type of guy. Short, stocky, well built. He used to tease my wife before we were married."

Fedol says the military crews that manned the reactor were young men of their time: playful, energetic, and full of vim and vigor. But he is quick to add that they had a healthy respect for the machine and possessed a maturity common to their generation. He didn't recognize the reckless images of Byrnes and Legg painted after the explosion by the AEC investigation. "When we were working, we were very serious young men," he says. "It was serious business, and we knew it."

Fedol says the managers of the plant—the army, Combustion Engineering, and the AEC—all knew SL-1 was experiencing serious problems with the control rods and the loss of boron but chose not to shut the reactor down. When the problems hit a critical level and it all went horribly wrong in less than a second that cold January night, he thinks the military and the AEC ducked and ran, leaving Byrnes and Legg as the scapegoats.

"I still believe in justice and fair play," says Fedol, who is disgusted that rumors about the two men still circulate four decades after the incident. "I would love to see these guys exonerated. No one has ever proven them guilty of anything. But they were assumed guilty."

Ed Vallario, the SL-1 health physicist who pulled the first body from the reactor on the night of January 3, remained a strong defender of the three crewmen throughout his life. Years after the accident, he returned to the Idaho site for a brief visit. When he heard a public relations person tell a group of visitors that the SL-1 explosion may have been caused by a love triangle, Vallario rebuked him. He was there. He knew the men. It was hogwash, he said.

"The rumors angered him," says his wife, Bette, also a health physicist. "They really upset him because he felt there was just no foundation in truth to them. This was a man who truly enjoyed people, and he thought those rumors were a great dishonor to both the men and their families. He thought it [the cause of the explosion] was a design flaw in the reactor."

Vallario's support of the three young men never wavered, nor did his belief that he did the right thing by plunging into the reactor to rescue the first crewman—even though it was an action that marked, and later doomed, him. "There was a residual red spot along the upper part of his chest near the collarbone that extended down three or four inches and was maybe as wide as two hands. I remember seeing it as a kid," recalls Robert Vallario, Ed's son, who himself went on to work in the nuclear field. "He said he never actually thought anything about it. But later on in life, many, many years later, he went for a medical exam and a doctor who was a little more astute said, 'What the heck is this red stuff on your chest—did you ever receive a burn or a radiation burn?' My dad finally put two and two together and said, 'It's interesting that you should say that...'"

Some years after that physical examination, and more than thirty years after he rushed into the SL-1 reactor, Ed Vallario was diagnosed with multiple myeloma, a cancer of the bone marrow. Both he and his doctor, says his wife, had no doubt that the relatively rare condition was the result of the gamma radiation that bombarded Vallario's body while he was inside the dank guts of the SL-1 reactor during the rescue mission.

Don Petersen, the radiation biologist who assisted in the autopsies of the three crewmen, believes critics of nuclear energy often overstate the danger of radioactive materials, particularly in small doses. In fact, in the mid-1950s, he often volunteered to ingest trace amounts of radioactive material for his friend Dr. Lushbaugh's early experiments in radiological medicine. "I've been accused of munching my way through the periodic table," jokes Petersen, still in good health more than forty-five years after the experiments. Still, trace amounts are one thing; the hellish cocktail of fission products in the SL-1 reactor shortly after the explosion another, he admits.

Petersen, who knew Ed Vallario, recognizes that the health physicist's dashes into the reactor may now seem rash, even foolhardy. But not on that January night. Then, they were heroic and emblematic of the mindset of the atomic pioneers in the Lost River Desert. "By the time Ed got to the reactor, they knew they had a problem and there were some victims," Petersen says. "They didn't know what kind of condition they were in. When you have that kind of ignorance and there's a possibility you could help, then you do some dumb things. Going in there and extracting those people like they did is probably something, in retrospect, that would be done differently had you known all the things you know after the fact."

Even a staunch defender of the industry like Petersen acknowledges Vallario's cancer was suspicious. "Multiple myeloma appears to be elevated among radiation workers," he says. "It's one of those things where it's ambiguous, but it's certainly something that has been questioned."

Throughout 1998, cancer ravaged Vallario, destroying his sculpted body and handsome face. His son remembers one hospital visit well, and in particular a comment his father made: "You know, when he—I hate to say it—was really sick, he allowed to me, 'I have to tell you based on the data that I have and what I know and with my background, this [illness] was work-related.'"

And what is the data? Vallario kept the records from readings off a dosimeter he wore into the reactor, as well as the results of a far more sophisticated radiation test he underwent soon after the accident. He also had the AEC's official record of his radiation exposure. The first two didn't even come close to matching the last. An emaciated, dying Vallario confided to his son that he believed the official documents were doctored to show he'd been exposed to far less radiation than he really had been.

After the accident, and to this day, the official story has been that twenty-two of the initial responders to the reactor explosion received radiation exposures ranging from three to twenty-seven roentgens of total body exposure. Three of the rescuers, the government maintains, received more than twenty-five roentgens of gamma radiation but no more than twenty-seven. The government still refuses to release the radiation doses absorbed by individual rescuers, but one of those three was presumably Vallario. At the time, the Idaho office of the AEC allowed rescue personnel to receive a one-hundred-roentgen dose to save a life and twenty-five roentgens to save valuable property.

Vallario's son Robert also believes his father's dose was far greater than what the government alleges it was. "The original dosimeter was greatly in excess of anything that was reported," he says. "He knew that, but he never said anything about it. His feeling was, 'Ah, I've got the records, but I understand what they did. And to me it's happened already. I did what I did because I had to. And irrespective of what dose I received, it wouldn't have changed what I was going to do. It had to be done, and I was the guy to do it.'"

Vallario was certainly in a position to blow the whistle: because of the radiation he had received that night, he was never again allowed to work in any proximity to a nuclear reactor. Not long after the Idaho disaster, he moved to Washington, DC, where he became the director of radiation protection for the AEC and later, for the Department of Energy. He helped write virtually all the radiation protection procedures imposed at government nuclear facilities, and he ended his career as chief of the Energy Department's health physics branch. More importantly, said his family, despite the nagging knowledge that he had absorbed a whopping amount of radiation, he lived life with curiosity and exuberance, rather than with fear.

"What do you do in life?" asks his wife Bette. "You can let it eat at you, or you can just go live. And my husband was the kind of person who just went out and did what he wanted to do in life."

"I don't want to make it [sound] too altruistic," says Robert Vallario of his father's experiences. "In this era, we approach things with such skepticism. But these guys had a very different reference point back then. They were like Chuck Yaeger; they were guys out in the Wild West—one minute they're up in the hills plinking with guns and the next they're jumping into hot reactors or flying X-15s at twice the speed of sound. He described it as a different time, you know? It was a time [when] we all lived under the threat of nuclear annihilation by the Soviet Union. What's judged by today's standards to be improper was, in those times, clearly thought to be in the national interest and in the best interest of the American public."

Robert insists that for his father, "not blowing the whistle was the right way to go...it was a sacrifice he was willing to make for the greater good. He actually said this to me in between stays at the hospital: 'If I had to do it all over again, I wouldn't have done it any other way.'"

A few years after the SL-1 disaster, Vallario and the six other men who dashed into the reactor that first night received the prestigious Carnegie Medal of Heroism, created in 1904 to recognize outstanding acts of selfless heroism. His son remembers that his dad put the medal on his treasured piano, though he was always modest when he talked about what he had done to earn it. The Carnegie medal was still sitting on the piano when Edward Vallario died on January 30, 1999.

Vallario's wife said her husband stayed in touch with many of the men who went into the reactor with him. Paul Duckworth, who helped Vallario rescue the first crewman, later died of cancer, as did health physicist Syd Cohen, who accompanied the two into the reactor on their second foray. Bette Vallario remembers that the cancers were of the type that could have been caused by radiation exposure. Still, she says, Duckworth and Cohen endured their illnesses with the same stoicism and dignity her husband had shown. Not one of them made a public fuss about their illnesses, filed a suit, or sought compensation when they discovered their night mission into the SL-1 reactor had come back to haunt them.

The family of AEC nurse Helen Leisen also believes that she was one of the uncounted victims of the SL-1 explosion. Leisen was in the back of the cramped ambulance that carried the first crewman taken out of the reactor and down to the checkpoint on the highway. Rescuers had thought he was still alive when they brought him out, and Leisen was applying mechanical artificial respiration while the ambulance sped toward an AEC doctor.

C. Wayne Bills, the head of the technical team, recalls that he was carrying a five-hundred-roentgen radiation detector when he threw open the door of the ambulance as it arrived at the checkpoint. The detector read a gamma radiation level of four hundred roentgens inside the close confines of the Pontiac station wagon.

"We had a fairly exact time [of how long] she was there in the ambulance, but she wasn't badged," he says, referring to the film badges cleanup crews later wore to record their exposure levels while working at the contaminated site. "A lot of these people weren't badged."

In addition to the exposure from the intense gamma rays coming off McKinley's body, the nurse, middle-aged at the time, picked up alpha and beta radiation on her white shoes and around her exposed ankles. She, too, died a few years later of cancer. Her family always believed that Leisen's death was the result of her contact with the soldier's body in the tight confines of the ambulance. Government files still contain black-and-white photographs of the woman's nursing shoes, which had been taken from her after it was found they'd been contaminated. They are a chilling memento of the atomic age.

Both Bills, a health physicist by training, and George Voelz, the then-medical director at the site, aren't quite sure what to make of the reported radiation exposures of the initial rescuers—nor the fate that befell several of them.

"Vallario and Duckworth's exposures were, you know, for emergency work and lifesaving purposes," says Voelz. "Their exposures were sort of within what had been outlined as reasonable for that kind of operation. But, yeah, we would have liked it if we hadn't had to deal with those high levels of radiation. One of the problems, of course, was at that time the radiation instrumentation used an upper limit of five hundred R per hour and the levels in the room exceeded that, so they were kind of going by the seat of their pants.

"For lifesaving purposes at that time, you could go up to one hundred R per hour, so that would be a fifth of an hour, or twelve minutes," he says. "It's all based on time if we assume the level in there was five hundred R an hour. But actually the levels in the room were higher than that—seven hundred or so as a general level."

Bills agrees that the radiation doses to Vallario and his fellow rescuers didn't seem outrageously high at the time, but concedes something inexplicable must have happened that night: "We knew [radiation] fields and we knew time facts, so we assigned a lot of them about twenty-five to twenty-seven [roentgens of total body exposure]. That certainly wouldn't be any sort of dose—I don't think—that would cause cancer.

But there may be something about fresh fission products we don't fully understand because down in Utah, where they got the clouds from the nuclear test bursts, they had a high incidence of cancer...If you look at what the probability is of Vallario and Syd Cohen and Duckworth and maybe Leisen getting cancer...that's pretty high."

And what of the fates of the twelve hundred people who helped in the months-long cleanup of the reactor and the contaminated site? Though hundreds of those workers were exposed to high penetrating radiation fields, the AEC reported that their time exposure had been kept within acceptable three-month limits. But the standards of acceptable exposure were repeatedly lowered in the years following the accident. There was never a long-term study of potential health effects, either among those first on the scene or those who later cleaned up the atomic mess.

Egon Lamprecht, the eager young firefighter who was on the team that responded to the reactor's alarm that night in 1961, is a grandfather now. He remains fiercely loyal to the nuclear industry and proud of the cutting-edge technology the Testing Station honed. But he remains bothered by how little was done to monitor the health of workers in the aftermath.

"After the initial response [of the firefighters], because of our exposure, they pretty well kept us the hell out of there. By today's standards, it would be a major disaster," he says. "This is my pet peeve with the Atomic Energy Commission. Initially, we peed in the bottle for about two weeks to see what was in the bottle and how it was flushed out. And they took stool samples. After that, all it was, was a typical fireman's annual physical, not emphasizing 'Hey, you guys got exposed. Let's see what's happening to you.' What got in my craw was, after the initial thing, it was swept under the table—'Yeah, he looks OK.'"

But biologist Petersen emphasizes that, in 1961, there was an entirely different set of priorities. "You have to understand our mindset back then," he says. "There was a sense of newness and excitement about what we were doing. There was no other reason that everybody worked seventy and eighty hours a week on jobs that didn't pay you any more than working forty. Back then, no one worried about what we considered low-level radiation. Everybody was excited about what we were doing and thought it was important. The Cold War was part of it. We were all worried about the Russians. If you go back and kind of put your feet in the shoes of that time, a different set of issues were important, and things that are regarded as important today, no one even thought about."

Ed Fedol remembers that era well. He had been working at SL-1 until two weeks before the explosion. He was back at Fort Belvoir, Virginia, the home of the Army Nuke program, when trainee McKinley's body arrived for burial at nearby Arlington National Cemetery. Fedol attended the funeral with a group of other soldiers and sailors in the nuke program; some were new trainees like McKinley had been, others were veterans who had worked at SL-1. After the short graveside service—with mourners kept at a safe distance—he and his military colleagues stood at rapt attention, listened to the haunting sound of taps, and snapped a salute at McKinley's flag-draped, gray lead box. He remembers that the cadre was "devastated, shattered" by the deaths of their mates. He also recollects the palpable sense of patriotism of that era, and of that day. Fedol says that as the Nukes walked back to their cars, they all had the feeling they had just said goodbye to a soldier who had simply died in an accident, not from a menacing technology that could not be controlled. Not one of the servicemen that day, he says, questioned their faith in the atom. And today, Fedol still doesn't.

"You know, nuclear power is still the most feasible and cost-effective thing to have," he says. "The problem is, what do you do with the waste? What do you do with all the spent fuel rods? What do you do with all the paper waste and the rags and the absorbing paper that you have to put in barrels and bury? It's a waste problem. You know, I've lived next door to a nuclear plant, and it's safer than a coal plant as far as I'm concerned."

* * *

Following the funeral at Arlington, Richard McKinley's wife, Caroline, took the couple's two children—John Michael, then three years old, and Ann Marie, nine months—back to Ohio, where she and Richard had married when McKinley was just twenty-three. Caroline and the fatherless kids disappeared from public view and were never subjected to the base rumors and sordid legends of SL-1 as the other two wives and their families were. Trainee Richard McKinley had been fresh, having arrived just three weeks before the explosion. He left the scene the same way he had entered it: the new guy, untainted.

Judy Legg wasn't as lucky as Caroline McKinley. After her husband's death, she moved back into her parents' home. On February 2, 1961, a month after Dick had been killed and less than two weeks after he had been buried, she gave birth to the couple's son, Michael Eugene. Judy's brother, Michael

Cole, was away on a Mormon mission when the newborn arrived. But he remembers his parents saying that Judy was devastated, suffered from postpartum blues, and felt uncomfortable being in Idaho.

"She recognized that she was a mother and she was alone," says Cole. "Her family had counseled her against getting married really early, and she just kind of decided on her own to do that. And I think she felt uncomfortable about that because the circumstances didn't work out very well."

Judy may have been experiencing a bout of guilt and remorse, too. She later told her older brother, without supplying details, that she'd been having second thoughts about her marriage to Dick Legg. "She obviously had struck it off with her first husband, but she wasn't as happy as she thought she'd be," he says. "I think that they had a lot of issues they were dealing with. And I really don't know much of that story, either. But she wasn't married that long either and, you know, it takes a while to get used to the idea."

By remaining in Idaho Falls, Judy had put herself at the epicenter of the early rumors about the cause of the SL-1 explosion. Her brother says Judy didn't talk to him about the incident except once, eight or nine months later, when Michael returned from his mission: "I remember that Judy mentioned that people thought the fellow on top of the reactor had committed suicide, that he deliberately pulled the rod. Judy mentioned that he was separated from his wife. But she said that in her opinion—and I'm giving you secondhand information here—he did not commit suicide. She was of the opinion that the rod was sticking and he was just working it and, for whatever the reason, the rod stuck and he pulled a little too hard."

Judy never discussed with her brother the other rumor: the love triangle. In fact, she addressed the issue only once, years later. But her flat-out denial of any involvement was the very thing that confirmed her as the woman in the rumor in some people's minds.

When Hanauer's memo about the love triangle finally broke in the press in 1979, it simply revealed the love-triangle angle; it didn't name or indict either the crewman or the wife allegedly involved. Among the few people who knew the main actors, however, speculation had always settled on Judy, if only because she was the one investigator Leo Miazga had questioned Idaho Falls residents about. When Hanauer's claim became public record, Ben Plastino, a longtime newspaper reporter for the *Idaho Falls Post-Register*, decided to investigate, hoping to determine the validity—or invalidity—of Hanauer's theory. Plastino, having covered Idaho Falls news for many years, had close ties to several people at the Testing Station. Working those connections, he looked into the "love affair" possibility and found "not an iota of evidence," he wrote years later in a small book on the history of the Testing Station. However, in a filmed interview just a couple of years after his investigation, he identified, albeit indirectly, Judy Legg as the rumored "other woman." Plastino revealed that he had spoken to the women purportedly mixed up in the affair: "I took the occasion to talk to the girl myself; I think she lives in Texas. She vigorously denied any such thing ever happening."

At that time, Judy Legg was living in Fort Worth, Texas— the only wife of the three crew members to live in that state. The general public would likely have been oblivious to that subtle identification. But among a small circle of insiders who knew the three crewmen and their wives, or who knew someone who did, Plastino's on-camera remark forever branded Judy as the femme fatale in the SL-1 love-triangle story. Her comment to Plastino would be her final word; she never spoke publicly after that one interview regarding the explosion that killed her husband and her supposed role in the human emotions at play on the night it happened.

Nature would deal Judy Legg—and Dick in the grave—one more dirty trick, although it would come from cells instead of atoms. Judy Legg had always been a vivacious woman, with a large circle of friends in Idaho Falls—a "delight," her brother says. Several months after the accident, her friends began dragging her out to social events, where she met a sailor, an instructor in the navy's submarine program at the Testing Station. Joe Brackney was older than Judy, being nearing thirty at the time, and an accomplished sailor; he'd been under the North Pole on the USS *Skate,* the first submarine to accomplish that feat. Michael Cole describes Brackney as a "gem, an absolute gem," and says that Judy fell madly in love with him. In September 1961, the two were married; it seemed that Judy had left the horrors of that January night eight months earlier behind her. Joe adopted Dick Legg's son, Michael Eugene, and treated him as his own. A year after Judy and Joe's fall wedding, the couple had a son of their own, and the four fell into the rhythms of happy family life. But a few years later, Judy and Joe began to notice that their oldest son, Dick Legg's biological son Michael Eugene, was different.

"He was born with some type of disorder," says Michael Cole, who suspects it was either autism or one of its related conditions. "He was able to count to, like, a billion when he was really small. He memorized every commercial he ever saw. He can tell you the history of baseball players and what their batting averages were, going back I don't know how many years. He's like an idiot savant."

Cole says Joe Brackney fiercely loved Dick Legg's son, but it became obvious to both him and Judy that they didn't have the resources to keep the boy at home. Michael Eugene was eventually institutionalized, and the couple struggled for years to pay the enormous bills. Over the years, military attorneys had contacted Judy several times, encouraging her not to sue over the explosion at SL-1. Confronted, however, with the cost of her son's care, she filed a lawsuit in a federal district court in 1979, claiming negligence led to the death of Dick Legg. She asked for one and half million dollars from Combustion Engineering and the University of Chicago, whose Argonne National Laboratory had designed the reactor. The suit was settled out of court for one million dollars. Michael Eugene's care is ongoing. According to Cole, Michael Eugene, now in his forties, is "not completely functional" and lives in an assisted-care facility in Texas.

Peering through a window into Judy's life might leave some thinking that life handed her an unfair share of grief. But Judy certainly didn't look at it that way, says Michael Cole. As she was getting on in years, Judy told her brother she was grateful that things had turned out as well as they did. Dick's death was tragic, but she was resilient and had found a soul mate in Joe Brackney.

Even toward the end of her life, Judy retained the wacky sense of humor her older brother found so endearing. "She told me once—I'll never forget—'I have two goals in life: I want to have a handicapped parking sticker and watch TV all day.'

"She eventually did get that sticker when she had to take Joe for dialysis," Cole says wistfully. "He had a severe problem with his diabetes, and toward the end of his life, dialysis was a constant thing. She'd wait the whole day until they got the thing done and then she'd drive him home. Her health wasn't good at the time, either."

Joe's body finally gave out in 1998. Judy died a year later. When her life came to an end, she was happy and fulfilled, her brother says. If Judy harbored secrets about SL-1—and her brother doesn't think she did—she took them with her.

Judy kept the story of SL-1 close to her. Four decades later, her second son, Scott, hadn't heard the part of the tale that most people find so alluring: "My parents have been virtually silent on the matter. I have not even heard any of the rumors. Murder-suicide? That's unusual. I don't think I've even heard that one."

Ironically, Judy's brother too eventually went to work at the Testing Station. Over the years, as people learned who he was, they'd ask—usually indirectly in an attempt to seem polite—about the rumors of a love triangle. Michael Cole's standard response was to point them toward the thousands of pages of government documents about the SL-1 explosion stored in government offices in Idaho Falls. He himself never visited the offices or opened the files: "The story is really interesting. I've thought it was interesting for years. But it was a little too close to home for me to dig into it."

Judy Legg and Arlene Byrnes didn't socialize in the same circles when they lived in Idaho Falls, says Arlene's friend Stella Davis. Judy was a local girl, without kids, "just different," Davis remembers. Arlene and Stella, on the other hand, shared an East Coast upbringing, had young families to keep them busy, and were perhaps simply interested in other things. But Judy and Arlene had one thing in common after the night of January 3, 1961: an aversion to talking about the reactor explosion.

For nearly twenty years after the SL-1 accident, Arlene had only to dodge personal questions from the occasional pesky newspaper reporter or the odd snoopy neighbor. But after the Hanauer memo hit the press in 1979, she and Judy were approached by writers, TV executives, and even the film industry. Both women knew what really fascinated people about the SL-1 story: the rumors of sex, murder, and suicide. Addressing those rumors, even if only to refute them, meant consenting to an intrusion into their privacy. Independent of each other, both women decided not to open up their first marriages, their intimate lives, and their thoughts to public scrutiny.

After the initial trauma of the explosion and the shock of being left alone with a child, Arlene adjusted to her new life back in New York. Her son adapted to growing up without a dad. He spent a lot of time with his paternal grandfather, and the two became close. When Jackie was nine years old, Arlene met and married an air force officer and moved to Nebraska. A second child, Charlotte, was born soon after. As the children grew older and Arlene reveled in her happy second marriage, the memory of the traumatic events in Idaho, once so laden with raw emotion, faded.

Talking publicly about SL-1 would have forced Arlene to confront questions about her first marriage—how troubled it was, how Jack might have reacted on the night of January 3. Arlene, after all, is perhaps the only person who has clear insight into Jack's state of mind in the minutes leading up to the explosion. That night, she later told friend Stella Davis, she had made two calls to Jack at the reactor: the first time, to talk about ending their marriage, and the second time, the call which didn't get through, to seek reconciliation. Those calls, especially the second—which may have quelled some of Jack's runaway emotions—were something Arlene just would not talk about. Or could not talk about, suggests her friend Stella Davis.

"No one wants to talk to you about it, right? I don't think she'll give you the answers," says Davis. "You're asking me if she was in love with him? Is that what you're saying? Of course. But it was just like every other marriage. You cannot show me a perfect marriage. Because if you say you're perfect, I'll say you're a big liar. And a marriage has its ups and downs. Some of us handle it one way and others handle it in another way. And Arlene has always been in self-denial about it all."

Arlene did bend her no-talk rule—but not by much—to briefly discuss her forty-plus years of silence and to defend the reputations of her late husband and Dick Legg. "My daughter, Charlotte, asked, 'How come, mom, the three of you women never did a book or a movie on that? Was there something else behind it?' And I said, 'You know what? There is absolutely no reason for it.' Us three women never talked after the accident—it just dropped. I would think one of us would have put out a book or a movie, but none of us wanted to pursue any of it—we wouldn't even talk about it. And there was no reason behind it, either. It's just one of those things. My thing was, you kind of have to watch out for your government."

Arlene also wasn't willing, then or now, to address the dark rumors that continue to plague SL-1 and its crew: "And it's because...you stay away from something because of all the stuff they were saying, all that stuff they were coming out with. There were rumors of joking [Dick Legg goosing Jack Byrnes]. There was all kinds of stuff going around. You just wanted to stay away from it."

Although Arlene would not agree to an extensive interview, she rejected the notion that her late husband or Dick Legg was somehow responsible for the world's first nuclear reactor deaths. She says that, despite a battery of evidence to the contrary, the two men were under no unusual pressure that night.

"They were young guys. They weren't stressed," she firmly contends. "We have more stresses in the workplace in today's world. Back then, living where nothing was around? No, there was no stress. Today's world has the stress. Back then, they didn't even think of it that way. There were just three young guys who wanted to work, and there was just a bad reactor. That's all.

"There were a lot of problems with that reactor which were covered up," Arlene says. "Nobody said anything about that, you know. There was all that behind the scenes, but it was very hard for us to get hold of [official documents] at the time because it was all very secret. Back then, they kept everything real quiet. You have to remember that. They didn't want us to know a whole lot."

Arlene confirms that all three wives eventually received settlements from the companies involved in the SL-1 reactor, with Judy Legg's by far the largest. As Arlene sees it, the decision to settle was a clear indication that those who designed and oversaw the reactor had no defense against a negligence claim.

"Judy got a lot more, I think, than any of the rest of us," Arlene remarks. "She settled for a million dollars. For something like that [the way Dick died], she should have gotten a lot more. All three of us should have. But we didn't. We just didn't. And for them to say that they [the crew members] were at fault even when you win your settlement...."

But why would the government implicate her husband? Why would investigators conclude that the disaster resulted from a crew member's "involuntary performance" or "malperformance"? Arlene's take on the situation cuts with a distrust she has honed for more than four decades.

"Why, they're crooked as crooked can be."

Today, Arlene lives in Colorado with her retired air force husband. The years since have been kind to her, and she's been blessed with love and a family, much like what she had dreamed of as she drove into Idaho's Snake River Valley for the first time in 1959. And there's still a reminder, a happy one, from those years: Jackie, her son by Jack, who was just shy of his third birthday when SL-1 exploded. Now John Byrnes is a forty-something systems analyst, husband, and father living in Nebraska. He has the same quick mind and inquisitive nature as his father.

John Byrnes has always been intrigued by what happened in the Lost River Desert that night in 1961. Since he was a kid, he has been sharing the story with anyone who will listen. He's fascinated with the technical details, the heroism of the rescuers, and the ingenuity they showed in responding to the unimaginable. He's also interested in the SL-1 story because his dad died on top of the reactor that night.

To John Byrnes, his father certainly isn't the cardboard cutout portrayed in government reports. But nor is he a fully remembered presence. John Byrnes has the perfect distance to feel wonderment of his dad being part of military and scientific history without the deep pangs of loss or anger over what people claim he did. When John was younger, it was just the right distance for school writing assignments. "I did my first report on SL-1 in the fourth grade," he says, "and I can't tell you how many times I recycled it over the years!"

Over the years, Byrnes collected many of the government documents on SL-1, including the final report by the investigation committee and the closely guarded autopsy report that details the horrific injuries his dad sustained. He even got his hands on his father's graded exam from reactor school; "He got very high markings."

When the Internet was first developed, Byrnes, who earned a university degree in mathematics, posted a Web page about the incident, along with pictures of the crewmen, information about the men who went into the reactor to help them, and even cutaway graphics of the central control rod.

What he didn't get, and wasn't even made aware of, were the reports by investigator Leo Miazga, the confidential memos that seemed to sully his father.

But it's doubtful those reports—a collection of anecdotes and hearsay, after all—would change his opinion about what happened to the SL-1 reactor. He's read the technical reports about the reactor's problems. He's talked to friends of the family back then. He's heard his mother's story. He believed in the fourth grade and he believes today that the explosion was simply an accident, and that his father, Dick Legg, and Richard McKinley died doing their duty. John Byrnes doesn't believe that there was a love triangle, and he doesn't believe that his father was unstable and pulled the rod to kill himself.

"My mother reassured me it was absolutely ridiculous that there was any hanky-panky going on," he says. "And any kind of mental health issue just doesn't seem consistent with my family history. My grandfather was a beautiful man, and my dad's two brothers have had happy, successful lives."

That's not to say he thinks his dad or his buddies were saints. But they were guys in their twenties with dreams, not worn-down men with nothing to lose. "I just have this picture of them being very typical army guys," Byrnes says. "At night they'd go out and drink Schlitz and party. These were very young men. But they were hoping to get ahead in an industry that was just blossoming. My dad's whole goal was to operate a nuclear plant on Lake Ontario."

The passage of years has allowed the adult John Byrnes to see how the night of January 3, 1961, negated one possible future but created another, much like it did for many of the main actors in the SL-1 saga.

"It sucks. My dad is dead. I grew up seven years without a dad," John says. "But I couldn't imagine any other life now. Who knows how different my life would have been if my father had lived and had come back to work at a reactor in upper state New York? I would never have met my wife, and I would never have had my children. It's really kind of bizarre if you think of all the implications as you go pinballing through life."

One of the implications of Jack Byrnes' early death, of course, is that he can forever be fixed in time in the minds of others. The son chooses to remember the father as he was captured on eight-millimeter family film. It was Yellowstone Park in October 1959, just before Jack Byrnes drove his wife and baby son into Idaho and the Lost River Desert. "He was somewhere in Yellowstone, standing next to his car. He was a slender, good-looking young man. He had blond hair and a crew cut," says John Byrnes. "You could see the dew glistening off the rocks."

Others, those who believe what the scratches on the control rod seem to indicate and who accept the picture of the troubled young man painted in Leo Miazga's reports, fix Jack Byrnes in history on January 1, 1961, at 9:01 P.M. He's on top of the SL-1 reactor. He's looking down at the head of Dick Legg and the port of the central control rod. His strong fingers are wrapped around the lifting tool. His knees are sinking slightly, and his muscles are tensing.

"Call it a moment of insanity," says George Voelz, the doctor who saw in the chemical plant's steel sink what the violence of the events of the next millisecond did to John's dad.

Epilogue

The story of America's first, and only, fatal nuclear reactor accident has never been—and ultimately can never be—told. There are things known and things unknowable. In the days following the explosion, the world's press descended on the remote Idaho Testing Station, producing a spate of "who, what, and where" articles. Reporters adequately sketched the mechanics of the explosion and the kind of cleanup that would follow. They listed the names of the victims, their survivors, and their hometowns. But missing in the stories was the cause, the why.

A year and a half later, the final report on the SL-1 incident was released by the Atomic Energy Commission. The history of the sticky rods and the loss of boron were detailed. The physics of the "nuclear excursion" were explained. Possible mechanical and chemical causes were explored and dismissed. The training and supervision of the crew were examined; the findings seemed damning to some, mild to others. In the end, all plausible explanations were dismissed but for one. The

AEC told Americans the explosion was caused when a crew-man pulled the central control rod too far too fast, vaporizing instantly the water in the reactor core and creating a steam hammer that blew the top of the reactor apart. But why would a crewman who had reconnected the central control rod at least four times before that January night make such a critical mistake? Human "malperformance" was all the report said. By whom? It didn't say. With investigative, autopsy, and radiation exposure reports sealed, the final report would have to stand as the final word on SL-1. The explosion was, in essence, declared a mystery that would never be solved. Bigger things were at stake for the nuclear industry, and it appeared officials were content to let the secrets of that January night lie undisturbed in the lead-wrapped coffins of Jack Byrnes, Dick Legg, and Richard McKinley. The AEC report was, after all, an admission that atomic energy and the people who created and harnessed it were not infallible.

Decades later, *Atomic Energy Insights,* a journal published by Adams Atomic Energies, Inc., devoted the July 1996 issue to SL-1 and its story's absence from the history books. In an article that probes the reasons for SL-1's obscurity, author Rod Adams reported that insiders, who did not want their names used, believed there were "unstated reasons for not releasing the [AEC] report. While the term 'cover-up' wasn't used, the phrase, 'Let sleeping dogs lie' was used more than once."

And the dogs did sleep. Years after the explosion, Fordham University's Natural Science Professor Eric J. Simon conducted an informal poll of students, professionals, and professors—including one professor with a graduate degree in nuclear engineering—and none of them knew that the world's first nuclear deaths occurred in Idaho. They, like most other people, mistakenly believed that the 1979 Three Mile Island incident was the first serious reactor accident in America.

In the aftermath of the explosion, the industry did take internal steps to prevent a recurrence of the SL-1 tragedy. Reactors around the country were shut down, while operating procedures were rewritten and emergency response plans strengthened. The AEC increased its supervision of nuclear contractors and reactor programs. Respirators were redesigned to prevent the fogging that hampered rescuers at SL-1. Detection meters were created to read much higher levels of radiation. No reactor would ever again be designed with the ability to go critical from the removal of one control rod.

But those were technical responses to the SL-1 incident. What of the human questions raised by the accident? Were undertrained, inexperienced operators let loose on a marginal reactor? Did a well-known prankster take a stunt with his coworker too far? Did a young guy with a host of young guy problems—a rocky relationship, money troubles, a fast-lane lifestyle—take his frustrations out on a delicate and dangerous piece of machinery? Did jealousy, infidelity, and rage burn even hotter than enriched uranium that night in the steel silo?

Those were—and are—uncomfortable questions for an industry that depends on everything working just right. Those are questions that can't be answered by better engineering, more sophisticated physics, and new materials. Those are questions, say insiders, that truly reveal the larger story behind SL-1, a story that is as pertinent to the nuclear industry and the public today as it was in 1961.

"What is the story? The story is, you can wreck one of these plants," says nuclear regulator Stephan Hanauer. "Something went very badly wrong, either in somebody's head or in some piece of machinery or in the execution of commands that never should have been given. So I think there's plenty of blame for everybody.... This technology really depends on people and machinery. I don't know if people really understand that or not. Both can make trouble."

George Voelz, medical director at the Testing Station at the time of the explosion, believes that the human factor in the SL-1 incident is still relevant, even with today's far larger and safer nuclear reactors. "I think the human element is present in many of these accidents," he says. "This reactor really has no relevance to [modern] power reactors because they're so much bigger and complex. If this was done as a deliberate act, you couldn't do anything like this in a power reactor now. But the general conclusion that human decisions or performance plays a role in many nuclear accidents is certainly true. In a high portion of accidents you'll find that if someone had done something differently, you wouldn't have had a problem."

Egon Lamprecht, the firefighter who initially responded to the alarm at SL-1, makes no claim of technical expertise. He doesn't pretend to understand the complex mix of technical problems, faulty design, and human failings that came together so badly that night in 1961. He doesn't pretend to be an expert on the implications the accident posed for the industry. But he was there. He's read the reports and talked to people who had first-hand knowledge about the crew and the reactor. And he's had decades to think about it.

"We will never know for sure if it was a murder-suicide or whether it was an accident," he says. "I know it was man-caused. No one can deny that. That control rod got pulled out by a human being.

"Why was it pulled? We'll never know. Dead men don't talk."

Appendix

Explosion Timeline

−500 milliseconds The central control rod withdrawal begins.

−120 milliseconds The reactor goes critical when the control rods reaches 16.7 inches; rod continues to its full 20-inch extension.

0 seconds The power of the nuclear excursion peaks at 19,000 megawatts; the fuel plates begin to vaporize as temperatures hit 3,740 degrees Fahrenheit.

0.5 milliseconds The nuclear energy release ends; the center fuel elements and central control rod blade and shroud are ejected from the core; the water column above the core begins to accelerate upward.

34 milliseconds The water column rushes into the lid of the vessel; shield plugs are ejected from the lid at speeds of 85 feet per second; the vessel rises out of its sheath.

160 milliseconds The first shield plug hit the reactor room ceiling; two-thirds of the water inside the reactor is expelled and 5 percent of the fission products are released.

800 milliseconds The reactor vessel hits the ceiling.

2,000–4,000 milliseconds The reactor vessel falls down and comes to rest in its sheath.

Sources

Author's note: Sources are generally attributed within the narrative of this book. Present tense attributions indicate the subject spoke with the author; past tense attributions indicate the material came from subjects' interviews with government investigators. Below, in alphabetical arrangement, are sources the author used—either directly or indirectly—to tell the story of the explosion of the SL-1 reactor.

Prologue

- *National Geographic* article, "You and the Obedient Atom," Allan C. Fisher Jr., September 1958
- *The SL-1 Accident*, a film produced the US Atomic Energy Commission's Idaho Operations Office, undated but circa 1963

Chapter 1: Nuclear Apprenticeship

- Cave Archeology of the Snake River Plain, Idaho, US Bureau of Land Management, December 1999
- Craters of the Moon, Historic Context Statements, National Park Service, August 1999

- *Coming of Age: Idaho Falls and the Idaho National Engineering Laboratory*, Ben J. Plastino, BookCrafters, Chelsea, MN, 1998
- *Idaho: A Bicentennial History*, F.R. Peterson, Norton, New York, NY, 1976
- *Idaho Falls, City of Destiny*, Mary Jane Fritzen, Bonneville County Historical Society, Idaho Falls, ID, 1991
- *Idaho Falls Post-Register*, Golden Jubilee Edition, September 10, 1934
- INEEL Comprehensive Facility and Land Use Plan, US Department of Energy, Idaho Operation's Office
- Interviews with John Byrnes, son of John (Jack) Byrnes, November 2001 and March 2002
- Interview with Martin Daly, graduate of "Army Nuke" program, April 2000
- Interview with Stella Davis, friend of Arlene Byrnes and wife of a former military supervisor at SL-1, November 2000
- Interviews with Ed Fedol, graduate of Army Nuke program and acquaintance of Legg and Byrnes, April and June 2000
- Interviews with Elwyn Legg, cousin of Richard Legg, November 2000 and January 2001
- *Proving the Principle: A History of the Idaho National Engineering and Environmental Laboratory*, Susan M. Stacy, US Department of Energy, Idaho Operations Office, Idaho Falls, ID, 2000
- *SL-1*, a docudrama, produced by Diane Orr and C. Larry Roberts, a Beecher Films/KUTV Inc. Production, Salt Lake City, UT, 1983
- Snake River Plain Volcanics, US Geologic Service, May 2001
- *The Story of Idaho*, VM Young, University of Idaho Press, 1990

Chapter 2: Atomic Energy Meets the Cold War

Much of the historical information about the National Reactor Testing Station, including the BORAX 1 and Air Force nuclear airplane projects, is drawn from the exhaustively researched *Proving the Principle: A History of the Idaho National Engineering and Environmental Laboratory* by Susan M. Stacy, US Department of Energy, Idaho Operations Office, Idaho Falls, ID, 2000

- *Coming of Age: Idaho Falls and the Idaho National Engineering Laboratory*, Ben J. Plastino, BookCrafters, Chelsea, MN, 1998
- Interview with Brad Bugger, employee of the Idaho National Engineering and Environmental Laboratory, August 2000
- Interviews with Egon Lamprecht, lifelong resident of southeast Idaho and a former firefighter at the National Reactor Testing Station who was among the first to arrive at the scene of the SL-1 explosion, June, August, and October 2000
- Interview with Homer Clary, longtime Idaho Falls resident, co-worker of John Byrnes at Texaco gas station, and acquaintance of both Byrnes and Richard Legg, March 2001
- Interview with Homer Clary, March 2001
- Interviews with Michael Cole, brother of Judy Legg and a former employee at the National Reactor Testing Station, October and November 2001
- Interview with Clay Condit, former civilian physicist assigned to the naval nuclear submarine project at the National Reactor Testing Station and later assigned to investigate the cause of the SL-1 explosion for the US Navy, August 2000
- Interview with Stella Davis, November 2000
- Interviews with Elwyn Legg, November 2000 and January 2001
- Interview with Melbourne Legg, cousin of Richard Legg, February 2001
- Interviews with Susan M. Stacy, historian and author of *Proving the Principle: A History of the Idaho National Engineering and Environmental Laboratory*, July and October 2000, January 2001
- *New York Times*, "Rickover Honored by AEC for Role in Submarine Program," January 18, 1961
- *Proving the Principle: A History of the Idaho National Engineering and Environmental Laboratory*, Susan M. Stacy, US Department of Energy, Idaho Operations Office, Idaho Falls, ID, 2000
- "SL-1 Accident," US Atomic Energy Commission Investigation Board Report, Joint Committee on Atomic Energy, Congress of the United States, June 1961
- *The Army's Nuclear Power Program: The Evolution of a Support Agency*, Lawrence H. Suid, Greenwood Publishing Group, Westport, CT, 1990

- US Atomic Energy Commission Investigation Board, transcripts of hearings held in January 1961, Idaho Falls, ID

Chapter 3: "There Must Be Something Wrong at SL-1"

The narrative of the personal and professional lives of John Byrnes and Richard Legg in the months leading up to the explosion is reconstructed using two confidential reports by Leo Miazga, investigator, Division of Inspections, US Atomic Energy Commission. The first, dated January 20, 1961, is addressed to EB Johnson and is titled "John A. Byrnes III, Idaho Nuclear Power Field Office, United States Army." The second report is dated July 25, 1962, and is titled"SL-1 Incident, Supplemental Report."

The statement that Byrnes accepted Mitzi's offer of sex is an educated assumption. The government has redacted a critical sentence in the "SL-1 Incident, Supplemental Report" that would answer the question definitively. However, a close reading of the report—and the fact that a portion of a critical sentence describing Byrnes's interaction with Mitzi is redacted—seems to buttress this assumption. Paraphrasing an interview with then army sergeant Paul Conlon, Miazga wrote: "Sgt. Conlon said that [Mitzi] proved to be a woman of easy virtue and suggested a price of $20 per person. He added that some discussion ensued and she reduced her price to $2 per person and that some of those present took advantage of her offer while others declined. He said to the best of his knowledge [redacted], while Legg declined." Conlon, through the Department of Energy's Idaho Operations Office, declined to comment. Two former high-ranking employees of the Testing Station who either read the unredacted report years ago or who talked to investigators declined to deny that Miazga had reported a liaison between Mitzi and Byrnes.

- Interview with Homer Clary, March 2001
- Interview with Stella Davis, November 2000
- "There was a buttoned up Mormon culture...": Interview with Department of Energy employee Julie Braun, a lifelong resident of the Idaho Falls area, August 2000

- US Atomic Energy Commission Investigation Board, transcripts of hearings held in January 1961, Idaho Falls, ID
- "SL-1 Accident," US Atomic Energy Commission Investigation Board Report, Joint Committee on Atomic Energy, Congress of the United States, June 1961
- "SL-1 Cadre Log," Department of Energy, Idaho Operations office records repository
- "SL-1 Reactor Accident, Interim Report," May 15, 1961, Combustion Engineering Inc.

Chapter 4: Wayward Atoms

- Confidential reports: "John A. Byrnes III, Idaho Nuclear Power Field Office, United States Army," January 20, 1961, and "SL-1 Incident, Supplemental Report," July 25, 1962, by Leo Miazga, Division of Inspections, US Atomic Energy Commission
- *Idaho Falls Post-Register*, community notices, January 3, 1961
- Interview with C. Wayne Bills, the US Atomic Energy Commission's deputy director of health and safety at the National Reactor Testing Station at the time of the incident, October 2000
- Interviews with Don Petersen, former radiation biologist at Los Alamos Laboratory and a member of the SL-1 autopsy team, October and November 2000
- Interview with Bette Vallario, wife of Edward Vallario, July 2000
- Interviews with Robert Vallario, son of Edward, September and October 2000
- Interview with Dr. George Voelz, US Atomic Energy Commission's director of medical services at the National Reactor Testing Station at the time of the incident, September 2000
- Interviews with Egon Lamprecht, June, August, and October 2000
- Interview with Stella Davis, November 2000
- Interview with Don Petersen, October and November 2000
- US Atomic Energy Commission Investigation Board, transcripts of hearings held in January 1961, Idaho Falls, ID; interviews with Edward Vallario, Walter Moshberger, Max Hobson, Paul Duckworth

Chapter 5: "Caution: Radioactive Materials"

- Interview with Vernon Barnes, former employee at the Testing Stations Chemical Processing Plant, May 2000
- Interviews with Elwyn Legg, November 2000 and January 2001
- Interviews with Don Petersen, October and November 2000
- Interview with C. Wayne Bills, October 2000
- Interview with Dr. George Voelz, September 2000
- *SL-1*, a docudrama, produced by Diane Orr and C. Larry Roberts, a Beecher Films/KUTV Inc. Production, Salt Lake City, UT, 1983
- "The SL-1 Reactor Accident, Autopsy Procedures and Results," C.C. Lushbaugh, D.F. Petersen, L.G. Chelius, T.L. Shipman, May 1961, Los Alamos Scientific Laboratory
- US Department of Energy, Human Radiation Studies: Remembering the Early Years (oral histories), Dr. Clarence Lushbaugh and Don Petersen
- US Atomic Energy Commission, "Interim Report on SL-1 Incident, January 3, 1961," The General Manager's Board of Investigation, January 27, 1961

Chapter 6: Accident Aftermath

- "Briefing on SL-1 Accident at NRTS," minutes of US Atomic Energy Commission Meeting No. 1687, January 11, 1961
- Curtis Nelson, cover letter, US Atomic Energy Commission, "Interim Report on SL-1 Incident, January 3, 1961," The General Manager's Board of Investigation, January 27, 1961
- *Deseret News and Telegram*, "A-Reactor Blast Kills 3 In Idaho," January 5, 1961
- Dr. Albert Heustis, commissioner of Michigan Department of Health, memo to Francis J. Weber, chief of the US Public Health Service, January 17, 1961
- "Incident at National Reactor Testing Station," memo from E.C. Anderson, chief of AEC's special projects branch, to US Surgeon General, January 4, 1961

- "Interment of Radioactive Remains," memo from Leon S. Monroe to superintendent of Arlington National Cemetery, January 1961
- Interview with C. Wayne Bills, October 2000
- Interview with J. Robb Brady, then editor and later publisher of the *Idaho Falls Post-Register*, August 2000
- Interview with Clay Condit, August 2000
- Interview with Richard Feil, former SL-1 cadre member, February 2000
- *Idaho Falls Post-Register*, "Three killed in Idaho Reactor Blast," January 4, 1961
- *Idaho Falls Post-Register*, "NRTS Blast Prompts Flurry of Phone Calls," January 5, 1961
- *Idaho Falls Post-Register*, "What Happened? The Question Is the Big One in Reactor Probe," January 6, 1961
- *Idaho Falls Post-Register*, "Probe in Explosion Persists at NRTS," January 7, 1961
- Memo from Frank K. Pittman, chairman of ad hoc committee on reactor recovery operations, to A.R. Luedecke, general manager, US Atomic Energy Commission, August 7, 1961
- "Monthly report on the SL-1 Incident, May 1961," memo from Ray Daniels, National Reactor Testing Station, to E.C. Anderson, chief of the US Atomic Energy Commission's special projects branch, June 5, 1961
- *Newsweek*, "Atoms: Hot stuff," February 27, 1961
- *New York Times*, "Fear Safety Question Will Produce Repercussions in Power Programs," January 4, 1961
- *New York Times*, "Reaction of AEC, Atomic Industry," January 5, 1961
- *New York Times*, "Probe On; 2 Bodies Recovered," January 6, 1961
- *New York Times*, "AEC Sees No Hazard Beyond Immediate Vicinity," January 9, 1961
- *New York Times*, "3rd Body Recovered," January 10, 1961
- *New York Times*, "AEC Lays January Explosion to Nuclear 'Runaway,'" January, 20, 1961

- *New York Times*, "AEC Criticizes Itself on Safety," June 11, 1961
- "Review of Available Information on the SL-1 Reactor Incident," a presentation to industry representatives, January 24, 1961, Germantown, MD, by Frank K. Pittman, director of the AEC division of reactor development
- *Scripps Howard News Service*, three-part series, by Marshall McNeil, appeared in various newspapers, including *the Albuquerque Tribune*, during the month of February 1961
- "Several months after the autopsies, the rumors prompted...": exchange of memos between Francis J. Weber, chief of the US Public Health Service, and George Drasich, president of Local 2-652, Oil, Chemical and Atomic Workers International Union, with attached report by local union member Donald Seifert, May 1961
- "SL-1 Reactor Accident, Interim Report," May 15, 1961, Combustion Engineering
- "Summary of Aerial Monitoring (January 4 to January 8)," memo from John Horan, director of health and safety, NRTS, January 8, 1961
- "Summaries of Calculated and Monitored Results of Contamination Following the SL-1 Incident at NRTS," memo to E.C. Anderson, chief of AEC's special projects branch, May 1961
- US Atomic Energy Commission, "Interim Report on SL-1 Incident, January 3, 1961," The General Manager's Board of Investigation, January 27, 1961
- "Unusual Activities of January 1961," memo from R.D. Coleman, public health service, NRTS, to E.C. Anderson, chief of AEC's special projects branch, February 20, 1961
- US Atomic Energy Commission Investigation Board, transcripts of hearings held in January 1961, Idaho Falls, ID; interviews with Paul Duckworth, Sergeant Robert Bishop, Robert Meyer, and various other SL-1 personnel

Chapter 7: Murder-Suicide?

- *Coming of Age: Idaho Falls and the Idaho National Engineering Laboratory*, Ben J. Plastino, BookCrafters, Chelsea, MN, 1998
- Confidential reports: "John A. Byrnes III, Idaho Nuclear Power Field Office, United States Army," January 20, 1961, and "SL-1 Incident, Supplemental Report," July 25, 1962, by Leo Miazga, Division of Inspections, Atomic Energy
- *Cult of the Atom: The Secret Papers of the Atomic Energy Commission*, Daniel F. Ford, Simon and Schuster, New York, NY, 1982
- "Documents Concerning SL-1 Incident to be Marked 'Official Use Only,'" memo from AEC's Idaho site manager Allan C. Johnson to division directors, January 16, 1961
- "Exposures Received in the Recovery of the SL-1 Reactor Site," memo from US Army Chemical Corps Radiological Unit, August 15, 1961
- Interview with C. Wayne Bills, October 2000
- Interview with Homer Clary, March 2001
- Interview with Richard Lewis, former military supervisor at SL-1, October 2000
- Interviews with Stephan Hanauer, Department of Energy employee and author of memo describing the incident as a murder-suicide, August and October 2000
- Interview with Dr. George Voelz, September 2000
- Interviews with Susan Stacy, July and October 2000, January 2001
- *SL-1*, a docudrama, produced by Diane Orr and C. Larry Roberts, a Beecher Films/KUTV Inc. Production, Salt Lake City, UT, 1983
- US Atomic Energy Commission Investigation Board, transcripts of hearings held in January 1961, Idaho Falls, ID

Chapter 8: Nuclear Legacy

- Annual Atomic Energy Commission Report to the US Congress, 1961

- *Associated Press,* "1961 reactor accident spawns $1.5 million suit," January 29, 1979
- *Coming of Age: Idaho Falls and the Idaho National Engineering Laboratory,* Ben J. Plastino, BookCrafters, Chelsea, MN, 1998
- "Exposures to Ionizing Radiation at SL-1," memo from G. Landon Feazell, US Army Chemical Center, September 8, 1961
- *Idaho Falls Post-Register,* "INEEL 50th Anniversary," various by-lines, June 2, 1999
- *Idaho Falls Post-Register,* "Idaho Energy Lab Cleanup at 1961 Nuclear-Accident Site," Jennifer Langston, April 7, 2000, transmitted by *Knight-Ridder News Service*
- *Idaho Falls Post-Register,* "Nuclear Accident Remains a Mystery," Brandon Loomis, exact date unavailable, 1996
- Interview with C. Wayne Bills, October 2000
- Interview with Arlene Byrnes, former wife of John (Jack) Byrnes, March 2001
- Interviews with John Byrnes, November 2001 and March 2002
- Interviews with Michael Cole, October and November 2001
- Interview with Clay Condit, August 2000
- Interview with Stella Davis, November 2000
- Interviews with Ed Fedol, April and June 2000
- Interviews with Stephan Hanauer, August and October 2000
- Interview with Bette Vallario, July 2000
- Interviews with Robert Vallario, September and October 2000
- Interview with Dr. George Voelz, September 2000
- *New York Times,* "7 Atom Specialists Cited By Hero Fund," October 12, 1962
- Record of Decision Abstracts, Idaho National Engineering Lab, December 5, 1991
- US Atomic Energy Commission, "Interim Report on SL-1 Incident, January 3, 1961," The General Manager's Board of Investigation, January 27, 1961

Epilogue

- Adams Atomic Energies, Inc., *Atomic Energy Insights*, Volume 2, Issue 4, Rod Adams, July 1996
- Interview with Arlene Byrnes, March 2001
- Interviews with John Byrnes, November 2001 and March 2002
- Interviews with Stephan Hanauer, August and October 2000
- Interviews with Egon Lamprecht, June, August, and October 2000
- Interview with Dr. George Voelz, September 2000
- Professor Eric J. Simon, Fordham University, Web page citation